PASTEUR

A BEGINNER'S GUIDE

DR PETER J GOSLING

Hodder & Stoughton

A MEMBER OF THE HODDER HEADLINE GROUP

Orders: please contact Bookpoint Ltd, 39 Milton Park, Abingdon, Oxon OX14 4TD. Telephone: (44) 01235 400400, Fax: (44) 01235 400500. Lines are open from 9.00–6.00, Monday to Saturday, with a 24-hour message answering service. Email address: orders@bookpoint.co.uk

British Library Cataloguing in Publication Data
A catalogue record for this title is available from The British Library

ISBN 0 340 79053 9

First published 2001
Impression number   10 9 8 7 6 5 4 3 2 1
Year   2005 2004 2003 2002 2001

Cover image supplied by Corbis.
Typeset by Transet Limited, Coventry, England.
Printed in Great Britain for Hodder & Stoughton Educational, a division of Hodder Headline Plc, 338 Euston Road, London NW1 3BH by Cox & Wyman, Reading, Berks.

Acknowledgements

I would like to thank my family for their support and encouragement during the preparation of this book. I am especially grateful to my parents, Bet and Jim, for providing a temporary haven in which the manuscript could be completed.

CONTENTS

Preface

There are few individuals in the history of mankind whose work has had such an impact upon the health and well being of others, as that of Louis Pasteur. Many books have been written about Pasteur and his discoveries. Some have been aimed at the complete novice while others have been written in more scientific terms for professionals with various levels of background scientific understanding. This book, however, is aimed at those who want to know about the life of the great man, his work and discoveries, and puts them in to context with the happenings of the day. It is a Beginner's Guide and it is hoped that the reader will wish to explore further within his or her chosen subject areas.

Pasteur – A Beginner's Guide is organized into 9 chapters, key facts and discoveries are highlighted throughout, and each chapter ends with a summary. The first two chapters introduce Pasteur the man, examining his life and work in its historical context. The third chapter considers the influence of the nineteenth-century religious climate, public health and advancing knowledge in medical science had upon Pasteur's life and works. Chapters 4, 5, 6, 7 and 8 look in more detail at his theories and discoveries, and the impact they had. The final chapter covers one of the legacies of Pasteur's work – the establishment of the Pasteur Institute.

Pasteur – A Beginner's Guide outlines the genius of the man. It provides his life story and an analysis of his work, the text places his achievements in context by looking at the technological, public health and historical context of the time.

An Introduction to Pasteur and his Work

WHO WAS PASTEUR?

Louis Pasteur was a French chemist who lived from 1822 to 1895. He was one of the greatest scientists of the nineteenth century, making fundamental contributions to both chemistry and biology. His work on the optical activity of tartaric acid, a waste product of wine-making, formed the basis of an important branch of chemistry called stereochemistry. This led him to study fermentation and his findings helped open up the science of the study of bacteria, called bacteriology. He established the process by which wine and beer could be preserved – this later became known as pasteurization. Pasteur was also instrumental in developing the Germ Theory of disease, which associated disease with germs, and the development of vaccines.

WHY WAS HIS WORK SO IMPORTANT?

Pasteur's first love was chemistry, but his career also led him to biology and medical science. In each of these fields he displayed amazing talent. The findings from his experiments form the basis of much of today's chemical and biological science. His name is probably most famous in regard to the preservation technique pasteurization, which although originally used by him to preserve wines and beers is now universally used to preserve milk and a wide variety of other food products.

PASTEUR'S CHARACTER

Pasteur was born at Dôle, near Dijon in the eastern region of France. He was the son of a tanner who was a veteran of Napoleon's Army, and Louis himself was intensely nationalistic. As a youth, Pasteur's chief interest and talent lay in painting, but his ambitions lay in a different direction. He decided that he wanted to go to Paris to get his education at the École Normale, France's leading educational institution.

He passed the entrance examination in 1842 but was disgusted with himself as he only came fourteenth amongst his group. Such was his pride, and perhaps vanity, that he delayed entrance for a year and re-entered the examination a second time, this time taking fourth place. This intense strength of will and an unflagging ambition were prominent traits of Pasteur's character and undoubtedly helped him in his dogged pursuit of work and career. However, this did not always make him easy to get on with. Staff and students found him authoritarian and inflexible. Later he was overpowering as an administrator; between 1857 and 1867, he was Director of Scientific Studies at the École Normale.

A BRIEF BREAKDOWN OF PASTEUR'S CAREER

Pasteur started his career at the École Normale in Paris and became deeply interested in chemistry, particularly in the science of crystals, crystallography. He studied a group of compounds called tartrates. It was known that a waste product of wine-making called tartaric acid had identical chemical compositions to a substance called racemic acid, but different physical properties. Pasteur discovered that the crystals of these compounds were irregular and that solutions could be produced which rotated light both to the left and the right. Pasteur's discoveries provided the basis for the branch of chemistry called stereochemistry. He concluded that such molecular irregularity fundamentally distinguished living from inanimate things. It was the properties of the living that subsequently fascinated him.

In 1854 Pasteur moved to the manufacturing centre of Lille, as Dean of the Faculty of Science. Here he began to study fermentation, including the souring of milk, the alcoholic fermentation of wine and beer, and the formation of vinegar. The predominant view of alcohol production by fermentation in the mid-1850s was of a purely chemical process in which sugar is transformed into alcohol. Within three years Pasteur had succeeded in showing that the souring of milk – or lactic acid fermentation – was in fact caused by a living microscopic, rod-like organism. When introduced into a solution of sugar containing a nitrogen source, such as ammonia, this organism caused the formation of lactic acid. He had shown that fermentation was dependent upon microscopic forms of life, with each fermenting medium serving as a food source for particular microorganisms.

Pasteur moved back to Paris in 1857 to become Director of Scientific Studies at the École Normale. Here he continued his work on germs and fermentation. He developed techniques for culturing germs in liquid broths, and he discovered that some bacteria were inactive in air while others were active only in the presence of air. He called the former anaerobic and the latter aerobic. These important discoveries helped in our understanding of the growth characteristics of bacteria. Pasteur's research programme, probing the specific actions of germs, blossomed into some of his most spectacular demonstrations of the relationships between germs, putrefaction and disease.

Continuing his work with the wine industry in mind, he solved the problem of why wines and beers sour with age. He proved that living germs were responsible for souring wine and that heating the wine to 55°C before leaving it to stand eliminated the problem. Later, he successfully applied the same principle (pasteurization) to beer and milk, marking a major step in the preservation of foods.

Pasteur turned his attention to the study of infectious diseases which he believed were *transmitted* by the germs that also caused them. He thought that only particular organisms could produce specific infectious diseases, and in the early 1870s discovered that parasites were responsible for a disease of silkworms; these moth larvae were of great commercial importance to the French silk industry. In 1878 Pasteur began to experiment with fowl cholera. It was by chance that, following a staff summer vacation, healthy birds were inoculated with 'stale' cultures of the cholera-causing germs. Pasteur was fascinated to discover that no serious disease followed the microbial challenge. Intrigued by this, he injected the same birds, and some others, with a different 'fresh' culture and was amazed to find that those that had been previously injected remained healthy, while the additional birds fell ill. He had shown that it was possible to *protect* chickens against fowl cholera. He had succeeded in immunizing the chickens with the weak, old bacterial culture. This afforded protection when he later gave them fresh, virulent samples. Pasteur dismissed the notion that this discovery happened by chance stating that 'chance favours the prepared mind'.

Pasteur went on to apply this technique – of inoculating weakened microorganisms as a vaccine to prevent infection – to other diseases. He successfully applied it to important diseases of his time, including anthrax. Anthrax is a highly contagious condition which commonly affects cattle and sometimes humans. Livestock losses due to infection with anthrax were immense in Pasteur's time. It was particularly ruinous because infectious anthrax spores remained in the fields from which infected animals had long been excluded. Having produced a vaccine, Pasteur famously staged a stunning public demonstration of its efficacy in an experiment. This was similar to the one he had used with fowl cholera, but employed farm animals, fresh broths of anthrax germs, and his vaccine. Such demonstrations gave additional credence to the Germ Theory, which linked living germs to infectious diseases.

Pasteur continued in his investigations into infectious diseases, endeavouring to solve problems, and contribute to prevention and cures. In 1850 he moved his attention to rabies, a disease dreaded since antiquity because its symptoms were so gruesome and death inescapable. Pasteur developed a vaccine for rabies using the infected spinal cords of rabbits, which had been dried to weaken the virulence of the rabies virus. He demonstrated the efficacy of his anti-rabies vaccine in 1884 in an experiment using dogs. Pasteur had found a way to give dogs immunity to rabies and he later showed that, because the incubation period of the disease was lengthy, vaccination worked even if the dogs had been infected for some time. This observation was put to good use when, a year later, a nine-year-old boy called Joseph Meister, who had been mauled by a dog thought to be rabid, was brought from Alsace to Pasteur's doorstep. All the indications were that the boy had only a few weeks to live before the onset of the dreadful symptoms of rabies. With nothing to lose, Pasteur began a course of therapy using his vaccine in a series of increasingly virulent and painful injections given over a period of a couple of weeks. It soon became clear that the symptoms of rabies were not going to appear. A second case was successfully treated three months later. These dramatic human-interest events captured the world's imagination and over the following 15 months, the vaccine was given to over 2000 people.

On a wave of national enthusiasm, the Pasteur Institute was set up in 1888 to treat patients with rabies and, by the end of the 1890s, over 20,000 people worldwide had been treated by Pasteur's anti-rabies procedure. When Pasteur died in 1895 he was buried in his Institute, which remains a centre of excellence for biological research.

✳ ✳ ✳ ✳SUMMARY ✳ ✳ ✳ ✳

- Pasteur was one of the greatest scientists of the nineteenth century, making fundamental contributions to both chemistry and biology.

- Pasteur's work on the optical activity of tartaric acid formed the basis of an important branch of chemistry, called 'stereochemistry'.

- His findings from studies on fermentation helped open up the science of the study of bacteria, called 'bacteriology'.

- He established the process by which wine and beer could be preserved which later became known as 'pasteurisation'.

- Pasteur was also instrumental in developing the germ theory of disease, which associated disease with germs, and the development of vaccines.

- Pasteur's research into the specific actions of germs, blossomed into some of his most spectacular demonstrations of the relationships between germs, putrefaction and disease.

2 Pasteur's Formative Years

BOYHOOD

Pasteur was born on 27 December 1822, at Dôle, near Dijon in the east of France. Most of his boyhood was spent 25 kilometres to the south-east at Arbois, amid the vineyard-covered slopes of the Jura Mountains. It was here that his poorly educated father, Jean Pasteur, kept a tannery. Pasteur was the descendant of generations of tanners, his great-grandfather having been an indentured labourer who had purchased his freedom. But Louis showed no interest in continuing the family profession.

THE ARTIST

As a youth, Pasteur showed little interest in anything but fishing and drawing. But, perhaps as an early indication of his tenacity and ability to excel in anything that he put his mind to, he moulded his skills to become a talented artist. He produced a number of accomplished pastel drawings, including portraits of his parents and friends. In fact, drawings produced by young Louis suggested that he could easily have become a superior portrait artist. His later drawings of college friends were so professional that Pasteur was listed in at least two compendia of nineteenth-century artists. But Pasteur's destiny was to lie elsewhere, and this became clear to him as his education progressed.

THE STUDENT

Pasteur's elementary education took place in Arbois where for 10 years, up until 1839, he attended primary and secondary schools. He was not an outstanding student during these years of education, still preferring fishing and drawing to other subjects. But, encouraged by his father, his desire to follow an artistic career was waning, as he became increasingly interested in chemistry and other scientific subjects.

His interest was soon to become an all-encompassing passion which drove his academic studies and was to lay the foundation of his remarkable scientific career. In 1839 he became a student at Besançon. He stayed for three years and earned his Bachelier ès Lettres (Bachelor of Arts) in 1840 and Bachelier ès Sciences (Bachelor of Science) at the Royal College in Besançon in 1842. The following year he gained entry to the most prestigious French educational establishment, the École Normale in Paris, which had been founded specifically to train outstanding students for university careers in science and letters. It was here that he became Licencié ès Sciences (Master of Science) in 1845, and, after acquiring an advanced degree in physical sciences, he won his Docteur ès Sciences (Doctor of Philosophy) in 1847.

A BRIEF BREAKDOWN OF PASTEUR'S PRIVATE LIFE

Pasteur's successes with his crystallography research did not go unnoticed amongst his peers. Three years after obtaining his doctorate he was appointed as Professor of Chemistry to the Faculty of Science in what was later to become the University of Strasbourg, in industrialized Alsace. It was while he was in Strasbourg that he met and married Marie Laurent, the daughter of a local headmaster. She became his devoted wife and two years later in 1851, she gave birth to their son, Jean-Baptiste Pasteur, who later became a diplomat.

Although Pasteur's career, fame and productive work continued at a pace, his health began to fail him and in October 1868, aged just 45 he suffered his first stroke. This left him partially paralyzed on his left side, but with typical determination he still managed to continue his ground-breaking work, providing a comprehensive analysis of two diseases of silkworms and their prevention. It was fortunate for the world that Pasteur's haemorrhage occurred on the right side of his brain, the side that actually controls the left side of the body. Had the stroke been on the other side, he would not only have had damage to the right side of his body, but may also have lost some or all of his ability to speak and to understand words, which would have brought his scientific career to an abrupt close. Happily, he continued his work unabated.

Shortly after his triumph with the rabies vaccine that saved the life of the badly bitten nine-year-old boy, Joseph Meister, in 1887 Pasteur suffered a second stroke. This time, sadly, his speech was affected and, although he lived and worked for a further seven years, his long creative period was at an end. In 1894 he suffered yet another stroke which, this time, left him totally paralyzed and he died on the 28 September 1895 at Château Villeneuve l'Étang, in Paris.

SCIENTIFIC VIEWS CHALLENGED BY PASTEUR

Before we can appreciate fully the magnitude of Pasteur's discoveries, we must put them in the context of the obscurity of the scientific views of the period. Also, we should not lose sight of the fact that scientific advances are usually made only after much controversy and deliberation. Evidence in support or rebuttal of a particular theory needs to be gathered and must be considered analytically before an acceptable thread of truth can be ascertained. Undoubtedly, one of the chief factors in Pasteur's success was his enthusiastic participation in debates on the scientific questions of the day.

During the mid-nineteenth century there were some fundamental aspects of biology and chemistry that remained unclear. It might seem incredible, but there was still much debate, for instance, as to whether or not life could be generated from non-living substances.

SPONTANEOUS GENERATION

The belief that living creatures arise either from inorganic elements (abiogenesis) or from the reconstitution of dead organic matter (heterogenesis), without any intervention by a living 'parent', was still held by many. This was known as the Theory of Spontaneous Generation. It may seem bizarre now, but the idea that beetles, eels, maggots and, following Pasteur's work, microbes could arise 'spontaneously' from putrefying matter was speculated on from Greek and Roman times. In the nineteenth century it still had not been experimentally proven one way or another. Many people thought that spontaneous generation provided an 'explanation' for the appearance of maggots on corpses and the like.

One of the strongest advocates of spontaneous generation in France was Felix-Archimède Pouchet (1800–72), Director of the Natural History Museum at Rouen. In 1859, Pouchet argued that there were three factors necessary for the generation of lower forms of life: decaying organic matter, air and water. He felt that light and electricity were also helpful. By 1860, the French Academy of Sciences was sufficiently interested in Pouchet's claims to offer a prize for contributions that would 'attempt, by means of well-devised experiments, to throw new light on the question of spontaneous generation'.

Against the advice of his colleagues, who saw dabbling in this field as thankless and unrewarding, Louis Pasteur took up the challenge with gusto. Although at that time a chemist, he had religious and philosophical reasons for disliking a doctrine which reduced life to chemical reactions. He was, after all , a devout Catholic and opponent of Darwin (1809–82) and his Theory of Evolution. Applying his analytical mind and with a hunger for finance to fund his research, he finally provided the evidence that disproved the centuries-old Theory of Spontaneous Generation.

Fermentation

The process of fermentation was also an area of controversy during the nineteenth century. The predominantly held view in the mid-1850s was that alcohol fermentation was a purely chemical process in which sugar is transformed into alcohol. To the German chemist Justus von Leibig (1803–73), who had put his great authority behind this view, the yeast cells that brewers used promoted fermentation through their death and decomposition. He believed that living, biologically-active yeast had nothing to do with the process. Leibig held this opinion despite the fact that Theodor Schwann (1810–82), the German microscopist who discovered animal cells in 1839, had already shown that yeast cells reproduce and grow during fermentation.

The yeast cells that were found in the fermenting vats of wine were recognized as being live organisms, but they were believed simply to be either a product of fermentation or catalytic agents which provided useful ingredients allowing fermentation to proceed. Those few biologists who earlier concluded that yeast was the cause of, and not the product of, fermentation were ridiculed by the scientific experts. The deep conviction of the somewhat arrogant scientific establishment of the day was that chemistry had come too far to allow a mysterious life force theory to challenge pure chemical explanations of molecular reaction.

Again Pasteur applied his considerable expertise to reveal the true role of living microorganisms in the process of fermentation, and this is considered in detail in Chapter 5.

PASTEUR'S GREATEST RIVAL
During scientific debate, many rivalries may develop. This will almost certainly occur in respect to the validity of theories as different scientists test them by experimentation.

Pasteur's greatest scientific rival was the German physician Robert Heinrich Hermann Koch (1843–1910). Koch, along of course with Pasteur, is considered to be the co-founder of the science of bacteriology. He discovered the tubercle bacillus in 1882 and the cholera bacillus the following year. He won the Nobel Prize for Physiology or Medicine in 1905 for his discoveries in regard to tuberculosis. But he is probably equally well known for the so-called 'Koch's Postulates', which he first formulated fully in 1890.

Koch's Postulates stated that:

* The organism must be found constantly in every case of the disease.

✹ It must be possible to cultivate the organism outside the body of the host in pure cultures for several generations.

✹ The organism, isolated and cultured through several generations, must be capable of reproducing the original disease in susceptible animals.

These were important 'rules' as they set down the conditions that should be met before an organism could be considered to be the causative agent of a disease.

Koch's fame, gained for his work on developing techniques for obtaining pure cultures of bacteria on agar plates and his discovery of the bacillus responsible for tuberculosis, brought him into open competition with Pasteur. This rivalry was exacerbated by Franco–German nationalism in the aftermath of the 1870 war.

✳ ✳ ✳ ✳SUMMARY ✳ ✳ ✳ ✳

- Pasteur was the descendant of generations of tanners but, as a youth, Pasteur showed little interest in anything but fishing and drawing.

- The young Louis Pasteur was a talented artist. His drawings suggested that he could easily have become a superior portrait artist.

- His interest in chemistry and other scientific subjects became an all-encompassing passion which drove his academic studies and was the foundation of his remarkable scientific career.

- Pasteur conducted a series of ingenious experiments that enabled him to dispel the centuries-old theory of spontaneous generation.

- Pasteur's greatest rival was the German physician Robert HH Koch, who is considered to be the co-founder with Pasteur of the science of bacteriology.

- One of the chief factors in Pasteur's success was his enthusiastic participation in debates on the scientific questions of the day.

3 Backdrop to Pasteur's Life and Work

To fully appreciate the genius of Pasteur we should consider the climate in which his life and work took place. It gives us an indication of some of the hurdles that Pasteur faced, in addition to the scientific challenges that he so skillfully and enthusiastically tackled.

THE RELIGIOUS CLIMATE

During the nineteenth century, the Church in France was going through a turbulent time and something of an upheaval. Catholic scientists and intellectuals, among them Pasteur, generally found it difficult to come to terms with the increasing secularization of knowledge. These difficulties were particularly acute in France, where Catholicism had ceased to be the official State religion. In order to restore the Catholic Church's authority, many French Catholics sought political and religious support from the Pope and the freedom to run their own Church schools in opposition to the State schools which had been established at the time of Napoleon. Although Church schools were allowed from 1850, State officials laid down the curriculum and inspected them. Perhaps not surprisingly, many Catholics became antagonistic towards the modern thinking professed by many scientists of the day. Likewise, anticlerical sentiments were rife among the more liberal French.

In 1863, Joseph Ernest Renan (1823–92), a French philosopher who had abandoned clerical training because of his growing scepticism, published *A Life of Jesus*, describing Christ as an ordinary man about whom legends had accumulated. A rift had formed between those who followed Renan and Darwin – believing that human reason could, through the application of scientific method, understand everything, a school of thought referred to as scientific naturalism – and those who felt that such a creed was intellectually impoverished. Catholic idealists considered science was impoverished because it was powerless to save souls and ultimately harmful to man because it threatened to destroy the ethical foundations of European civilization. Scientific naturalism was, to many, irresponsible. Other critics pointed to the growing amount of vivisection which lay at the heart of the new physiology, and which the devout Pasteur was also criticized for supporting.

The anti-vivisection movement began in the 1880s in response to the development of experimental physiology, pharmacology and vaccine therapy. While Pasteur had no moral objection to vivisection – the preparation of the rabies vaccine involved live rabbits and that for smallpox required young calves – he usually made his assistants carry out such operations. The anti-vivisection controversy led to legislation which permitted experimentation to be carried out, but only under licence.

The French physiologist Charles Richet (1850–1935), who was to win a Nobel Prize in 1913 for work on serum therapy, championed the concept of scientific naturalism. Richet, who was a keen critical investigator of psychic phenomena, however, was prepared to admit that science did not have an answer to everything and that scientists did not always care about morality. However, he could not help but add that looking for the truth, even without regard for the consequences, was good in itself since it dispelled human ignorance! The French chemist Pierre Berthelot (1827–1907) also placed his faith firmly in a future in which science would continue to transform the material and moral conditions of human existence. Catholic scientists including Pasteur and the physicist Pierre Duhem (1861–1916) responded by founding the Catholic Scientific Society of Brussels and an international journal dedicated to the premise that knowledge of God was as certain as that of the physical world.

In 1870 the Franco–German War intervened. This conflict was opened by a declaration of war by Napoleon III but the Germans, who were better prepared than the French, won victory after victory. On 12 September 1871, Napoleon and 104,000 of his men were made prisoners at Sedan. A republic was then proclaimed, and Paris sustained a four months' siege. In the end, France conceded Alsace and part of Lorraine to Germany, which claimed a war indemnity of £200 million. The war had led to the foundation of a French Third Republic dedicated to the cult of science, and Catholic scientists were frequently snubbed. A point in case occurred in 1892 when the Republican government created the world's first chair in the History of Science but appointed an elderly rationalist rather than the better-qualified Catholic historian of mathematics, Paul Tannery (1843–1904). The latter was again deliberately overlooked in 1903 when the chair became vacant.

NINETEENTH-CENTURY MEDICINE

Until the beginning of the nineteenth century, the development of medical knowledge took place in two ways: firstly by the elaboration of systems for classifying symptoms and, secondly, through speculation about the nature of pathology, often attributing all disorders to a single cause treatable by a single remedy or set of remedies. The physicians in this period were encouraged to work in this fashion by the nature of their relationship with their patients. The physicians were the servants of wealthy patrons, who had a keen amateur interest in medicine. Since physicians did not have an effective legal monopoly on the provision of medical treatment, they were forced to compete with each other and with droves of irregular healers to attract patients. The emphasis on deriving medical theories from ancient Greek or Roman authors, for example, reflected the prestige attached to that learning by the educated classes among whom the physicians worked.

Since neither of these rival systems of medical knowledge had more than a limited success, patients' main criterion for judging their treatment was the extent to which they believed that they had received any benefit from their physician's work. The doctor's manner and his ability to impress patients, therefore, had a great influence on his prospects of employment. Many of the blood-curdling remedies of this period were accepted by patients because of the way they symbolically dramatized the physician's healing power. Many doctors placed great stress on the niceties of courtly behaviour, cultivating their wit, dress and style in order to encourage patients to identify and value them as gentlemen whose competence and reputation could be assumed from these outward signs.

This style of work first began to diminish in France after the Revolution of 1789. Until then, the provision of hospitals had been dominated by the Catholic Church. Before effective treatments were available, a good deal of medical care inevitably involved ministering to the spiritual comfort of patients for whom little else could be done. Consequently, medical care available in hospitals, for instance, was a matter of concern to religious interests. The close association between the hospitals and the Church meant that they were initially closed by the Revolutionary Government as part of its attack on organized Conservative opposition.

But, just as the English State had found in the sixteenth century, the new French regime was forced to recognize that the suppression of Church-based systems of care and welfare left a vacuum which had to be filled by secular equivalents, if the government were to command popular support. The revolutionaries reopened the hospitals under State ownership as part of a general plan to expand health and welfare facilities for the poor. This policy continued after the restoration of the Monarchy and, by the time Pasteur was born in 1822, the hospitals of Paris were caring for about 15,000 patients at any one time, some five times as many as were in all the voluntary hospitals of England and Wales.

Doctors in France now had access for the first time to very large numbers of poor people who lacked the power or resources to resist doctors' all too often fatal role in the advance of medical knowledge. The mortality rates were easily twice those quoted for English voluntary hospitals. Many of the surgeons, in particular, had developed their skills during the revolutionary wars. Under lay control, experimental medicine could be practised in the French hospitals without the old constraints of Catholic moral teaching.

Such operations were not without their critics, even in Paris, but their tragic results for individual patients paved the way for major advances in therapeutic technique and the understanding of human biology. Of equal importance was the free availability of corpses for dissection. At a time when the English medical colleges only had access to the bodies of a handful of executed criminals each year, their French counterparts in public hospitals had the right to dissect all cases terminating fatally under their care, unless the relatives could pay immediately for a funeral. As a result, diagnoses in life could readily be compared with signs *post mortem*.

In the first decades of the nineteenth century, doctors flocked to Paris in order to learn from this new style of medicine. There were estimated to be 200 English students of anatomy alone in the city in 1828.

Medicine, then, slowly began to become an occupation based upon a scientific model of research, rather than on ancient learning and armchair speculation. This new style of practice brought conflicts with traditional institutions. The doctors wanted to use hospitals as locations for studying collections of bodies, living or dead, organized so as to group like cases and differentiate unlike. Admission would be based on the value of a patient's condition for teaching or research purposes rather than on subscribers' patronage. The process of reorganization went on throughout the nineteenth century.

In its initial stages, the new style of medicine had placed great emphasis on the direct investigation of the patient by the doctor in order to identify and chart the course of the underlying disease. As the classification of diseases became more clearly defined, however, medical interest shifted towards using this knowledge in a rational search for better treatments. Pasteur, though not a physician, made fundamental advances in disease treatments and preventative medicine.

Treatments for infections

There was almost total ignorance of the nature of infection until the pioneering studies of Pasteur and Koch in the middle of the nineteenth century. Before that time, remedies largely centred around various plants and herbs, the use of inorganic salts and a variety of remedial 'tortures', such as bleeding, caultery, etc. Major advances in chemical knowledge occurred at the end of the eighteenth and beginning of the nineteenth centuries, and the newer techniques allowed the isolation of purer drugs from natural sources. Drugs such as aconite, belladonna, opium, morphine, strychnine, atropine and quinine were available by the middle of the nineteenth century, although none of these was used to treat infections.

Pasteurian bacteriology opened up the vision of biological, as distinct from chemical, agents being used to destroy bacteria. The first clear observation of antibacterial action was made in 1877 by Pasteur who noted that, while anthrax bacilli rapidly multiplied in sterile urine, the addition of 'common bacteria' halted their development. But it would be many years before the first antibiotic, penicillin, a natural by-product from moulds of the genus *Penicillum*, was to be described by Alexander Fleming (1888–1955).

OUTBREAKS OF DISEASE

In the mid-nineteenth century life was pretty desperate, particularly for the working population. Poverty and physical misery were widespread, and so were sickness and early death. But infectious diseases affected all classes. Pandemics, epidemics covering a wide geographical area (usually many countries), of disease cast regular shadows over many parts of the world during the 1880s. France was no exception.

Cholera

Pasteur was ten years old when the first of several deadly epidemics of cholera ravaged France. Part of a worldwide pandemic that started in India in 1826, cholera advanced relentlessly westward across Eastern

Europe in the early 1830s. France and other countries helplessly awaited its arrival. Having no experience with cholera, a disease new to Europe, French health officials were sent to already stricken foreign cities to study preventative measures and medical treatments. Although most members of the medical establishment did not endorse the idea of contagion, the government's Council of Health insisted on quarantine measures.

In Paris, the disease came to public notice on 29 March 1832. The speed with which symptoms overtook the victims, who often died within hours, caused immediate and widespread panic. The sporadic and unexpected course the disease seemed to take appeared to be a mystery. One area of the city would be ravaged while another remained untouched, or one side of a street would be stricken and the other spared. Those who supported the theory of contagion were hard-pressed to explain these haphazard appearances in terms of direct transmission from one infected person to another, while anticontagionists felt that cholera's haphazard progress corroborated their belief that contaminated air (miasma) engenders and diffuses disease. Despite the debates about cholera's origins, most medical discussion focused on treatment. But all approaches proved completely useless. As in times of plague, people who could afford to do so fled the cities and towns to escape infection. As many as 120,000 people left Paris almost at once, and 10,000 left Marseilles in January 1833. No doubt, some found refuge in the rural area of the Jura Mountains, where Pasteur still lived.

Unlike childhood diseases, such as smallpox and measles, which were considered a sad but inevitable fact of life, or influenza, which was usually fatal to the elderly, cholera killed as many healthy young adults as people of other age groups – a characteristic it shared with plague. An estimated 100,000 people died from cholera in France during the 1832–3 epidemic. With its harrowing symptoms, high mortality rate of 25 per cent to 50 per cent, and its ability to kill people in the prime of life, cholera caused more terror than any disease in European history except plague.

Pasteur was to witness wave upon wave of cholera outbreaks through his life. Destructive and widespread outbreaks followed the country's first experience of the disease. These occurred in 1848–9, 1853–4, and 1865–6, while less serious occurrences erupted in 1873, 1884 and again in 1892.

French Miliary Fever

French Miliary Fever accompanied France's first outbreaks of cholera, and it was very widespread in 1841–2 in the south-west, where it killed some 30,000 people. In 1850 epidemics occurred in pockets near the German and Swiss borders, and in 1860 in the French province of Burgundy. In March 1887, it broke out in Montmorillon in western France and spread quickly through the surrounding area. The mortality rate during this epidemic ranged from 9 to 25 per cent, depending on the locality.

Smallpox

Pasteur was 48 years old at the outbreak of the Franco–German war (1870–1), which unknown to the world, was to trigger a devastating epidemic of smallpox. Up until this time, smallpox cases in Europe had been gradually declining in number and severity because of the introduction of crude vaccination programmes, both voluntary and compulsory, in several countries. However, in France, smallpox had been festering silently, and the death rate due to the disease had increased about sevenfold since the mid-1860s. War preparations, troop movements, and the migration of Parisian citizens out of their city exacerbated the previously isolated smallpox outbreaks in France. The disease began to spread rapidly, killing between 60,000 and 90,000 people in France in 1870 and 1871. The epidemic was not limited to France and Germany, however, and spread throughout Europe, killing at least half a million Europeans.

Influenza

During his lifetime, Pasteur will have also witnessed the devastating effects of wave upon wave of influenza. France was affected by pandemics sweeping Europe in 1830–1, 1836–7, 1847–8, and 1889–90, and suffered high infection and mortality rates. More deaths worldwide were caused by influenza than by any other disease in the nineteenth century.

Against this backdrop of incessant epidemics sweeping through France during the nineteenth century it is perhaps not surprising that Pasteur turned his attention to treating and preventing infectious disease in humans.

DEVELOPING MEDICAL SCIENCE

During the eighteenth century, observation and experiment became the driving forces of contemporary advances in the physical and natural sciences, building on the earlier emphasis of the power of reason and natural, rather than supernatural, explanations for phenomena. This led to an expansion of what became known as 'scientific methods' of investigation, which aimed to discover genuine and practically useful knowledge, drawn from observation and stated in terms of universal laws.

The application of this approach to medicine resulted, during the nineteenth century, in the rapid development of medical science. Increasingly, emphasis was placed upon discovering universal laws governing the causes of disease by means of the 'clinico-pathological method'. This involved correlating symptoms with abnormalities in the structure or function of part of the body, discovered either at *post-mortem* or in the living patient. During the nineteenth century, the medical notion of disease had moved entirely away from seeing it as an affliction of the whole body, to seeing diseases as highly localized disturbances of anatomical structure.

By observation of the facts of numerous particular instances, recurrent patterns of symptoms and lesions could be identified. This revealed the regularity of nature in the development of specific diseases, such as anthrax, to which Pasteur contributed invaluable knowledge. This resulted in doctors gaining the power to *predict* future outcomes when similar facts were presented. The laws apparently governing the development of pathological phenomena could now be identified.

In this new climate, professionally trained doctors were anxious to ensure that their practice was securely founded on a scientific basis. This meant recognition of the biological basis of the complaint, as well as the laws of its development.

* * * *SUMMARY * * * *

- Catholic scientists and intellectuals found it difficult to come to terms with the secularization of knowledge.

- The anti-vivisection movement began in the 1880s in response to the development of experimental physiology, pharmacology and vaccine therapy.

- The Franco–German war (1870–1) led to the foundation of a French Third Republic dedicated to the cult of science.

- Doctors in France had access for the first time to very large numbers of poor people who lacked the power or resources to resist doctors' all too often fatal role in the advance of medical knowledge.

- Medicine slowly began to become an occupation based upon a scientific model of research, rather than on ancient learning and armchair speculation.

- Pasteur made fundamental advances in disease treatments and preventative medicine.

- Major advances in chemical knowledge occurred at the end of the eighteenth and beginning of the nineteenth centuries, and the new techniques allowed the isolation of purer drugs from natural sources.

- Pasteurian bacteriology opened up the vision of biological agents being used to destroy bacteria.

- Pasteur was ten years old when the first of several deadly epidemics of cholera ravaged France.

- Pasteur was 48 years old at the outbreak of the Franco–German war (1870–1) which, unknown to the world, was to trigger a devastating epidemic of smallpox.

- During his lifetime Pasteur witnessed the devastating effects of influenza.

A New Branch of Chemistry – Crystallography

It may surprise some readers to learn that Pasteur, the 'father' of bacteriology and immunology, was a chemist who launched his memorable scientific career by studying the shapes of organic crystals.

Long before 1822 when Pasteur was born, French science had already established a tradition of research into the nature and properties of crystals. Crystallography is the name now given to the study of the structure – both internal ionic and molecular arrangements – and morphology of crystals, and their classification into different types.

From the 1770s onwards, mineralogists had been particularly active and one of their number, René Just Haüy (1743–1822), is regarded as the founder of the science of crystallography through his discovery of the Geometrical Law of Crystallization. He theorized that the molecules of different minerals show limited different basic forms, and that it was how they were joined together that produced different gross structures. He, therefore, thought that there was a finite number of different primitive forms from which all crystals could be derived by being linked in different ways. Using his theory, he was able to predict, in many cases, the correct angles of the crystal face. The findings from such studies on the structure of molecules touched on many different scientific disciplines.

By 1815, a French physicist, Jean Baptiste Biot (1774–1862), had observed that many organic liquids were, like some crystals, capable of rotating the direction of polarization of a beam of polarized light that was passed through them. He noted that the beam emerged at an angle to the plane of vibration of the original polarized beam. Since these organic molecules were not crystalline, Biot concluded that their power to alter the light must be inherent in the structure of their molecules.

During Pasteur's lifetime, another French scientist, this time a chemist called Auguste Laurent (1807–53), had also become interested in crystallography. His interest lay in the use of an instrument called a goniometer. This was a device for measuring angles and bearings, and he considered it could be used as an instrument of analysis to identify substances crystallographically by their chemical forms. In 1846, having recently returned to Paris from Bordeaux where he was Professor of Chemistry, he worked for a short time in Professor Antoine-Jerôme Balard's laboratory. It was fortuitous that, while there, he met Louis Pasteur, then a 26-year-old student, and suggested that Pasteur should investigate the relationship between optical activity and crystalline form.

Pasteur considered whether this would be a suitable area of study and became fascinated by crystallography, which was then just emerging as a separate branch of chemistry. He began to crystallize a number of different compounds, and started working with tartaric acid.

Tartaric acid is a normal constituent of grapes and had long been known to form the principal ingredient of the sludge which separates from wine during fermentation. It was known since 1844 that there were two kinds, or isomers, of the organic compound tartaric acid. An isomer is one of a series of compounds that have the same molecular formula; that is, they contain exactly the same number of atoms of every element. But in tartaric acid one isomer, the active form, turned the plane of transmitted polarized light to the right. The other, the inactive form, apparently produced no effect at all. An optical isomer is called an enantiomer. But since their crystalline shapes and, therefore by implication, their molecular forms seemed to be identical, it was a puzzle as to how their solutions could have such different effects on polarized light. This is the problem that Pasteur tackled successfully for his doctorate in 1847.

THE IMPORTANCE OF PASTEUR'S WORK

Having prepared samples of crystallized tartaric acid, Pasteur carefully examined his yield. Intensely detailed comparisons made with a magnifying glass showed that the inactive crystals were symmetrical, whereas the active crystals were asymmetrical. The polarization effect seemed, therefore, to depend upon the existence of asymmetric crystals. Upon preparing larger optically inactive crystals, Pasteur noticed by eye that there were, in roughly equal proportions, two different kinds of crystals in the sample. Both appeared asymmetric, but the crystals were mirror images of one another, like right- and left-hand gloves.

Intrigued by this, he then performed one of the simplest and yet most elegant experiments in the annals of chemistry. With a dissecting needle and his microscope, he painstakingly separated the left- and right-hand crystal shapes from each other to form two piles of crystals. Next, he made solutions of the two types of crystals he had separated. He showed that one form of crystal rotated light to the left, now known as the laevo form, and the other to the right, now known as the dextro form. Pasteur concluded that, since one type rotated the plane of polarization to the right, and the other rotated it to the left, and each were present in approximately equal proportions in the inactive sample, polarization effects were normally cancelled out. That is the form of tartaric acid that appeared inactive was in reality made up of a mixture of equal amounts of asymmetrical crystals.

Pasteur had prepared a new form, the laevo form, of tartaric acid which did not occur naturally in isolation from the dextro form. He may be considered fortunate in his choice of crystal to study, as only ten or so crystalline racemates – the name given to substances made up of equal quantities of both enantiomers – can be separated optically. What is more, tartaric acid racemate will only separate into dextro and laevo forms below 27°C. However fortunate his discovery was, it was to form the basis of an important branch of chemistry, called stereochemistry.

DEXTRO OR LAEVO?

While at Strasbourg, Pasteur asked himself why nature appears to be 'handed'. Why do naturally occurring crystals always uniquely possess either dextro or laevo form? Pasteur speculated that there was an asymmetric force present in nature – perhaps light, heat, or electricity. He experimented with the effects of giant magnets on chemical reactions to see if optically-active, rather than optically-neutral, crystals were produced. They were not.

In 1854 he arranged for plants to be grown in sunlight whose rays were reversed by means of an optical device, to see whether they produced dextro rather than laevo compounds in their leaves. The results were inconclusive. In any case, Pasteur saw that there was no experimental way he could remove the influence of any symmetric materials that were already present in the plant seeds themselves. (DNA, deoxyribonucleic acid, which we know now to be the key molecule in the genetic structure of living systems, is, in fact, dextro-rotatory.) It seems that the 'handedness' of the compound, both in nature and in the laboratory, can only be achieved through the intervention of an optically active intermediate in the synthesis.

PASTEUR'S MOVE TO BIOLOGY

During his investigations Pasteur tested the effect of fermentation upon solutions of tartrates. Using both the dextro and racemic forms of tartaric acid, he found that fermentation brought about laevo-rotatory activity in the latter. In other words, during fermentation, the dextro-tartaric acid in the neutral racemic acid was consumed by some living organism, leaving behind only the laevo-tartaric acid. Not only did this give Pasteur a new method of resolving mirror-image isomers, but it revealed that they were different physiologically. Bacteria would feed upon the dextro-acid, but not on the laevo form.

On the basis of these experiments, Pasteur elaborated his theory of molecular asymmetry, showing that the biological properties of chemical substances depend, not only on the nature of the atoms constituting their molecules, but also on the manner in which these atoms are arranged in space. He had formulated a fundamental law: that asymmetry differentiates the organic world from the mineral world. In other words, asymmetric molecules are always the product of life forces. His work became the basis of a new science called stereochemistry.

To Pasteur these discoveries had a deeper meaning. He proposed that asymmetrical molecules were indicative of living processes, and we now know that this, in the broadest sense, is correct. We know today that all of the proteins of higher animals are made up of only those amino acids that exist in the left-hand form. The mirror image right-hand amino acids are not used by human or animal cells. Likewise, our cells burn only the right-handed form of sugar, not the left-handed form that can be made in the test tube.

It was the discovery of asymmetry in organic molecules that led Pasteur into the study of fermentation as a chemical and biological process, and slowly shifted his career from chemistry to biology.

✳ ✳ ✳ ✳ SUMMARY ✳ ✳ ✳ ✳

- Pasteur made solutions of the two types of crystals he had separated from tartaric acid and showed that one form of crystal rotated light to the left, now known as the laevo form, and the other to the right, now known as the dextro form.

- Using both the dextro and racemic forms of tartaric acid, he found that fermentation brought about laevo-rotatory activity in the latter. During fermentation, the dextro-tartaric acid in the neutral racemic acid was consumed by some living organism, leaving behind only the laevo-tartaric acid.

- Bacteria fed upon the dextro-acid, but not on the laevo form.

- Pasteur elaborated his theory of molecular asymmetry, showing that the biological properties of chemical substances depend not only on the nature of the atoms constituting their molecules but also on the manner in which these atoms are arranged in space. He noted that asymmetry differentiated the organic world from the mineral world.

- The discovery of asymmetry in organic molecules that led Pasteur into the study of fermentation as a chemical and biological process, and slowly shifted his career from chemistry to biology.

Fermentation and Pasteurization

In 1854 Pasteur moved to Lille, another French industrial centre, as Dean and Professor of Chemistry of the Faculty of Science. Faculties were special schools for higher education which Napoleon had established in the provinces in 1802 as part of his educational reforms. Since one of Lille's leading industries was the fermentation of beetroot and grain to make alcohol, and as a logical progression from his successful studies with tartaric acid, Pasteur commenced his own studies in and offered courses on fermentation.

WHAT IS FERMENTATION?

Fermentation is a term used today to describe the series of enzyme-catalyzed reactions occurring under anaerobic conditions in certain cells, particularly yeasts. During this process organic compounds such as glucose are converted into simpler substances with the release of energy. In alcoholic fermentation a process known as glycolysis occurs which, under anaerobic conditions, converts glucose into the alcohol ethanol and the gas carbon dioxide.

During the time of Pasteur's work on fermentation, however, the process was still shrouded in mystery. Chemistry was just emerging as a true science, freed from the pseudo-science of the alchemist. The mysterious chemical processes of living animals were slowly being unravelled but strictly in chemical terms. In this light, fermentation leading to production of wine, beer and vinegar was believed to be a straightforward chemical breakdown of sugar into the desired molecules. The chemical experts of the day proclaimed that the breakdown of sugar into alcohol during the fermentation of sugar to wine and beer was due to the presence of inherent unstabilizing vibrations. One could transfer these unstabilizing vibrations from a vat of finished wine to fresh grape pressings to start a new round of fermentation. Yeast cells were found in the fermenting vats of wine, and were recognized as being live organisms, but they were believed simply to be either a product of fermentation or catalytic agents that provided useful ingredients allowing fermentation to proceed.

PASTEUR'S BIOLOGICAL VIEW OF FERMENTATION

Although, as a chemist, Pasteur might have been expected to have shared Leibig's 'pro-chemical' view of fermentation, his experience of the asymmetric nature of fermentation products inclined him to a biological interpretation.

By 1857 he had succeeded in showing that the souring of milk, or lactic acid fermentation, was caused by a microscopic rod-like plant which, when introduced into a solution of sugar containing a nitrogen source such as ammonia, caused the formation of lactic acid. Pasteur had demonstrated, as had Schwann before him, that fermentation was dependent upon the presence of microscopic forms of life, with each fermenting medium serving as a unique food for a specific microorganism. Although Liebig was shown to be in error in this exchange, it was later realized that yeast and other microorganisms promote fermentation by the secretion of cell chemicals called ferments, or enzymes. The compromise between these two extreme views was one of the factors in the development of a new science, biochemistry, in the 1880s.

Pasteur's intense nationalism also played a part in his interest in brewing; German beers and wines were strong opposition to the French products. He was also much struck by the continuous nature of the yeast's growth during alcohol fermentation, and did not know of any analogy with other chemical reactions. Upon visiting factories and breweries Pasteur became aware that batches of beers and wines were commonly spoiled causing serious economic problems. Yields of alcohol might suddenly fall off; wine might unexpectedly grow gelatinous or sour or turn to vinegar; vinegar, when desired, might not be formed and lactic acid might appear in its place; the quality and taste of beer might unexpectedly change, making quality control a nightmare! All too often, the producers would be forced to throw out the spoiled batches and start afresh, sometimes with no better success.

Pasteur quickly found three clues which allowed him to solve the puzzle of alcoholic fermentation. He noted that when alcohol was produced normally the yeast cells were plump and budding. But when lactic acid formed instead of alcohol small-rod like microbes were always mixed with the yeast cells. Chemical analysis of the batches of alcohol showed that amyl alcohol and other complex organic compounds were being formed during the fermentation. This could not be explained by the simple catalytic breakdown of sugar. Some additional processes must be involved.

The critical clue to Pasteur was that some of these complex organic compounds rotated light; that is they were asymmetric. As Pasteur's work with crystallography had shown, only living cells produced asymmetrical compounds. He concluded and was able to prove that living cells, the yeast, were responsible for forming alcohol from sugar, and that contaminating microorganisms turned the fermentations sour.

WHAT IS PASTEURIZATION?

Over the next several years Pasteur identified and isolated the specific microorganisms responsible for normal and abnormal fermentations in production of wine, beer, and vinegar. He demonstrated that abnormal fermentation of wine and beer could be prevented by heating the beverages to about 57°C (135°F) for a few minutes. This killed living microorganisms, thereby sterilizing (pasteurizing) the batches and preventing their degradation. If pure cultures of microbes and yeasts were added to sterile mashes uniform, predictable fermentations would follow.

The same pasteurizing effect is realized by heating to 72°C for 15 seconds.

THE IMPORTANCE TO INDUSTRY

Today pasteurization is a major strategy in preserving food and preventing or delaying food spoilage. Pasteurization normally involves heating food to a relatively low temperature for a short time. The time and temperature combination chosen depends on the particular type of food, and must be sufficient to destroy vegetative pathogens and a considerable proportion of spoilage organisms. Toxins and spores generally survive pasteurization and, to avoid the growth of heat-resistant organisms and vegetative bacteria which may be present because of spore germination during heating, refrigerated storage is often necessary.

Pasteurization of milk is, in particular, widely practised in several countries. This requires temperatures of about 63°C (145°F) to be maintained for 30 minutes or, alternatively, heating to a higher temperature of 72°C (162°F) for 15 seconds, or yet higher temperatures for shorter periods of time. The times and temperatures are those determined to be necessary to destroy the *Mycobacterium* species and the other more heat-resistant of the non-spore-forming, disease-causing microorganisms that may be found in milk. The treatment also destroys most of the microorganisms that cause milk spoilage and so prolongs the storage time.

The main advantage of pasteurization is that food is rendered safe with the minimum effect on flavour and nutritional value. Modern foods which are commonly pasteurized, include milk, ice cream, liquid egg, wines, canned fruit and large cans of ham.

Pasteur's work on fermentation helped open up another new science, bacteriology.

✳ ✳ ✳ ✳SUMMARY ✳ ✳ ✳ ✳

- Fermentation is a term used to describe the series of enzyme-catalyzed reactions occurring under anaerobic conditions in certain cells, particularly yeasts. During this process organic compounds such as glucose are converted into simpler substances with the release of energy.

- In alcoholic fermentation a process known as glycolysis occurs which, under anaerobic conditions, converts glucose into the alcohol ethanol and the gas carbon dioxide.

- Pasteur demonstrated that fermentation was dependent upon the presence of microscopic forms of life, with each fermenting medium serving as a food source for particular microorganisms.

- Pasteur was able to prove that living yeast cells were responsible for forming alcohol from sugar, and that contaminating microorganisms turned the fermentations sour.

- Pasteur demonstrated that abnormal fermentation of wine and beer could be prevented by heating the beverages to about 57°C (135°F) for a few minutes. This killed living microorganisms, thereby sterilizing (pasteurizing), the batches and preventing their degradation.

- The main advantage of pasteurization is that food is rendered safe with the minimum effect on flavour and nutritional value.

A New Science – Bacteriology

WHAT ARE BACTERIA?

Bacteria are tiny living things which can be seen under a fairly powerful light microscope but not by the naked eye. Some are rod shaped (bacilli), others grouped like bunches of grapes (staphylococci) or in strings or chains (streptococci). Yet others are comma shaped, such as the cholera vibrio, or shaped like a drumstick, such as the tetanus bacillus which causes lockjaw and has the appearance of a rod with a small knob at one end.

It would be a mistake to think that all bacteria are harmful, for without some species we could not survive for long. Medical bacteriologists divide these organisms into three groups according to their behaviour in the human body: saprophytic, parasitic or pathogenic, and symbiotic.

* The *saprophytic* organisms are the bacteria normally found on the skin, and in the mouth and intestines; they do us neither harm nor good.

* The *parasitic* or, as they are more usually called, *pathogenic* (i.e. disease-producing) organisms are the harmful ones with which we are naturally more concerned.

* Lastly, there are the *symbiotic* organisms, which, whilst taking something from the body, give something in return. For example, cattle would not be able to digest the cellulose of the grass they eat were it not for the helpful bacteria in the lower parts of their intestines, and there are certain bacteria in the large intestine of humans which produce vitamins.

Unlike bacteria, viruses, as Pasteur discovered when experimenting with the rabies virus, are too small to be seen under an ordinary microscope. Today some viruses can, however, be photographed under an electron microscope, which uses a magnetic field instead of a glass lens and a steam of electrons in place of a beam of light. Viruses cause such diseases as measles, mumps, poliomyelitis, smallpox and chicken pox, not to mention such plant and animal diseases as tobacco mosaic disease and foot-and-mouth disease, both of which can have serious economic consequences for farmers. The main characteristics of viruses are, first, that they can only grow in living cells. This is in contrast to bacteria, which readily grow in the laboratory on plates containing a nutrient jelly. Secondly, many viruses are so small that they pass through the pores of the finest filter. Thirdly, a first attack usually produces immunity in the sufferer for life. Second attacks of the common viral diseases mentioned above are very rare; but, unfortunately, this rule does not apply to influenza or the common cold.

Some infections are caused by fungi, organisms belonging to the same group as moulds, mushrooms, and toadstools. Penicillin and some other antibiotics are produced by moulds so, as in the case of bacteria, some fungi are helpful; they even help to destroy each other, as bacteria do. For example, the fungus actinomyces, which can cause infection of the jaw and other tissues, is destroyed by penicillin. Most fungal infections are trivial and limited to the skin. But, although trivial, they can be unsightly and uncomfortable. Ringworm of the scalp and so-called 'athlete's foot' are also caused by a fungus. Other fungi, such as the yeasts, are used in the brewing industry to ferment wines and beers.

Parasites may live on the skin. Examples include lice, which carry typhus, fleas, carriers of plague, and the parasites of scabies which burrow into the skin. Other parasites may live part of the time in the blood or other tissues, like the malaria parasite. These organisms often have complicated life-cycles involving other hosts, like mosquitoes, at certain stages of development. Pasteur's work with silkworms brought him in contact with parasitic infections.

THE BEGINNING OF BACTERIOLOGY

The beginning of bacteriology began with the rejection of the Theory of Spontaneous Generation (see pages 56–57). Based on his work on fermentation, it seemed obvious to Pasteur that the yeasts and other microorganisms which were found in the products of fermentation and putrefaction entered from the outside, for example, on the dust of the air. Pasteur conducted a series of ingenious experiments that destroyed every argument supporting spontaneous generation. He showed that the skins of grapes were the source of the yeast. Drawing grape juice from under the skin with sterile needles gave juice that would not ferment. Wrapping developing grapes in cotton to keep off contaminating dust gave grapes that would not produce wine. In order to show that dust in the air was the carrier of contamination, he allowed air collected at different altitudes – from sea level to mountain tops – to enter sterilized vessels containing fermentable solutions.

The higher the altitude, the less the dust in the air, and the fewer flasks showed growth. The experimental design that clinched the argument was the use of the swan-neck flask. In this experiment, fermentable juice was placed in a flask and, after sterilization, the neck of the flask was heated and drawn out as a thin tube with a gentle downward then upward arc, resembling the neck of a swan. The end of the neck was then sealed. As long as it was sealed, the contents remained unchanged. If the flask was opened by nipping off the end of the neck, air entered but dust was trapped on the wet walls of the neck. Under this condition, the fluid would remain sterile, showing that air alone could not trigger growth of microorganisms. If, however, the flask was tipped to allow the sterile liquid to touch the contaminated walls and this liquid was then returned to the broth, growth of microorganisms immediately began.

Pasteur considered that the doctrine of spontaneous generation would never recover from the mortal blow of this simple experiment. But the arguments continued for many years, complicated by the religious and ideological aspects, and it was not until the mid-1870s that the theory was discarded once and for all. On the positive side, however, the skirmishes over spontaneous generation resulted in development of the Germ Theory and the study of bacteria, creating the new science of bacteriology.

WHAT IS BACTERIOLOGY?

Bacteriology is the name now given to the branch of microbiology dealing with the study of bacteria. The beginnings of bacteriology paralleled the development of the microscope. Although others may have seen microbes before him, Antonie van Leeuwenhoek (1632–1723), a Dutch draper whose hobby was lens grinding and microscope making, was the first to provide proper descriptions of his observations. These included protozoans from rainwater and the guts of animals, and organisms that seem to correspond with some of the very large forms of bacteria we now recognize, from teeth scrapings and other substances.

Leeuwenhoek's original microscopes had, at best, a highest magnification of 266 and resolution of 2 micrometres. Despite this, his descriptions and drawings were excellent because his lenses were of an exceptional quality. Leeuwenhoek conveyed his findings in a series of letters to the British Royal Society during the mid-1670s. Although his observations stimulated much interest, no one made a serious attempt either to repeat or to extend them. Leeuwenhoek's 'animalcules', as he called them, thus remained mere oddities of nature to the scientists of his day, and enthusiasm for the study of microbes gained ground only slowly.

As late as the mid-nineteenth century, bacteria were known to just a few experts and only in a few forms, mainly as curiosities of the microscope, and chiefly interesting for their minuteness and motility. Modern understanding of the forms of bacteria originates from the brilliant classifications of the German botanist Ferdinand Julius Cohn (1828–98).

Cohn, stimulated by the work of Louis Pasteur, had become increasingly interested in bacteria. His classic treatise *Researches on Bacteria*, published in 1872, laid the foundations of modern bacteriology. In it he defined bacteria, used the constancy of their external form to divide them into four groups. This widely accepted classification was the first systematic attempt to classify bacteria and its fundamental divisions are still used in today's nomenclature.

Modern bacteriological techniques had their beginnings in 1870–85 with the introduction of the use of stains and the discovery of the method of separating mixtures of organisms on plates of nutrient media solidified with gelatine or agar. Important discoveries came in 1880 and 1881, when Pasteur succeeded in immunizing animals against two diseases caused by bacteria. His research led to a study of disease prevention and the treatment of disease by vaccines and immune serums (a branch of medicine now called immunology). Still other scientists recognized the importance of bacteria in agriculture and the dairy industry.

Bacteriological study subsequently developed a number of specializations, among which are agricultural, or soil, bacteriology; clinical diagnostic bacteriology; industrial bacteriology; marine bacteriology; public health bacteriology; sanitary, or hygienic, bacteriology; and systematic bacteriology, which deals with taxonomy.

PASTEUR'S ROLE IN THE FOUNDATION OF BACTERIOLOGY

While Cohn and others advanced knowledge of the morphology of bacteria, Louis Pasteur and other researchers established the connections between bacteria and the processes of fermentation and disease, on the way discarding the Theory of Spontaneous Generation and improving antisepsis in medical treatment.

It was in Paris that Pasteur developed techniques for culturing microbes in liquid broths and he discovered that some bacteria were *inactive* in air or, more specifically, in oxygen. Others were only active in the presence of oxygen. He called the former anaerobic, the latter aerobic.

The identification of anaerobic bacteria, which obtain their oxygen indirectly from other compounds, suggested to Pasteur that decay or putrefaction, like fermentation, was another biological process. Anaerobic bacteria began the processes of fermentation and putrefaction, and the products they produced were then acted on by aerobic bacteria. One consequence of this was that microorganisms were of great significance, for without them the world would become saturated with undegraded organic molecules. There would be no recycling of materials.

It was to demonstrate this point that Pasteur designed a classical experiment. In the first of two flasks he placed a solution of yeast with some preheated sugar or milk and pure dust-free air; the second flask contained the yeast, unheated sugar or milk and unpurified air. Both flasks were kept for several days at a temperature of about 30°C, following which the air in both flasks was analyzed. In the first flask, the air still contained a large quantity of oxygen; but in the second, where the access of microscopic organisms had been unimpeded, the oxygen was exhausted and replaced by carbon dioxide. Clearly fermentation, whether in the anaerobic or aerobic mode, only proceeded in the presence of microorganisms.

CULTURING BACTERIA

Pasteur's technique for cultivating pure homogeneous strains of bacteria was to use a liquid growth medium, such as sugar solution laced with yeast ash. Once growth had begun, Pasteur used a minute amount to inoculate a fresh medium, and so on, until he was personally satisfied by microscopic analysis that he had a pure culture.

In about 1880, a commercial meat extract (a beef broth) developed by Justus von Liebig was found to be an ideal nutrient medium. However, by then it had become clear that, despite Pasteur's impressive achievements using a liquid medium, pure cultures were not guaranteed and that misleading results were being produced. Another of Pasteur's techniques, breeding a pure culture through 20 or more generations of mice or some other experimental animal, had the disadvantage of being slow.

It was Pasteur's greatest rival, Robert Koch, who, on analyzing in 1880 the ideal conditions for culturing bacteria, saw that the medium ought to be sterile or sterilizable, transparent in order to show the growth more clearly and, above all, solid, so that mixing of bacterial species could not occur. Koch's first solution, in 1881, was the humble potato, followed by a meat extract in gelatine which could be made to set on sterile glass plates, or as a drop on a microscope slide. These were inoculated in a 'noughts and crosses' pattern and cultured at a uniform temperature under a bell jar. If more than one species of bacteria developed they appeared first as distinct colonies before merging into one another. By sampling and re-inoculating a clean plate before merging occurred, Koch obtained a pure culture of an organism.

This culturing technique was demonstrated by Koch at an international medical congress in London in 1881 and it rapidly became a routine procedure. The following year, the wife of one of his assistants suggested that agar; an extract of seaweed should be used instead of gelatine, which many bacteria can break down. The familiar petri dish, with its compact lid, was devised by an assistant, Richard Petri, in 1887. This technology of culture set the Germ Theory on a firm foundation.

✳ ✳ ✳ ✳SUMMARY ✳ ✳ ✳ ✳

- Bacteria are tiny living things which can be seen under a fairly powerful light microscope, but not by the naked eye.

- It would be a mistake to think that all bacteria are harmful, for without some species we could not survive for long.

- Unlike bacteria, viruses are too small to be seen under an ordinary microscope.

- The beginning of bacteriology began with the rejection of the Theory of Spontaneous Generation.

- Bacteriology is the name now given to the branch of microbiology dealing with the study of bacteria.

- Modern bacteriological techniques had their beginnings in 1870–85 with the introduction of the use of stains and the discovery of the method of separating mixtures of organisms on plates of nutrient media solidified with gelatine or agar.

- Pasteur and other researchers established the connections between bacteria and the processes of fermentation and disease, on the way discarding the Theory of Spontaneous Generation and improving antisepsis in medical treatment.

- Pasteur developed techniques for culturing microbes in liquid broths and he discovered that some bacteria were inactive in air, or more specifically, in oxygen. Others were only active in the presence of oxygen. He called the former anaerobic, the latter aerobic.

- Fermentation, whether in the anaerobic or aerobic mode, only proceeded in the presence of microorganisms.

7 The Germ Theory

WHAT IS THE GERM THEORY?

In the mid-nineteenth century many aspects of disease remained a mystery. At that time the most widely held theory of disease was the Miasmatic Theory. This theory associated disease, and epidemics like cholera and influenza, with poisonous fumes given off from dung heaps and decaying matter. These fumes, or 'miasma', were blown by prevailing winds from one area to another and this was thought to explain how epidemics spread.

A better understanding of microorganisms and their role in disease and various 'chemical' processes brought forth an alternative theory; the Germ Theory. This alternative theory came about once Pasteur had shown that decay was due to microorganisms, and not the often obnoxious vapours they produced as a *by-product* of their actions. It became possible to picture germs as organisms which 'germinated' in animal and human hosts, and to regard diseases as an interference with the normal chemistry of the body by a 'parasitic' invasion.

In 1878 Pasteur argued the case for the Germ Theory of disease before the French Academy of Medicine. He spelt out his conviction that microorganisms were responsible for disease, putrefaction and fermentation; that only particular organisms could produce specific disease conditions; and that, once those organisms were recognized, defensive strategies would be possible.

Development of the Germ Theory was an inspirational leap forward in our understanding of the role of microorganisms in infectious disease. It provided the basis for the development of vaccines and other efforts directed toward the prevention of disease, either in the individual or in the community as a whole, an important part of what is now broadly termed 'public health'.

WHAT WAS PASTEUR'S CONTRIBUTION?

Pasteur's contribution to the establishment of the Germ Theory began through his practical work on the 'diseases' of wine and beer. In 1862 he investigated the manufacture of vinegar (acetic acid) from wine. He showed that the scum on a vinegar vat consisted of rod-like microorganisms which promoted the oxidation of the alcohol in the wine to form acetic acid. The presence of these organisms helped explain the souring of wine and led manufacturers to consult Pasteur about other malfunctions of wine-making and brewing. In all cases, he found that the yeasts employed to cause fermentation had become contaminated with foreign organisms. This is why, in 1865, Pasteur recommended the heat treatment of wines and beers to 55°C before leaving them to stand.

DISEASE OF SILKWORMS

Pasteur also contributed to the Germ Theory through an elegant study into the cause of disease in silkworms. This may seem an unlikely subject of study for a chemist actively studying fermentation. But his successes and involvement in the spontaneous generation debate, together with his propensity for publicity, had afforded him considerable notoriety. In 1865 he was asked by the chemist and former Minister of Agriculture, JB Dumas, to help solve a serious problem in the French silk industry. During the 1860s the industry was almost destroyed by a disease of silkworms which made their eggs sterile or left the larvae unable to feed. Between 1850 and 1866 French silk producers had lost nearly 20 million francs.

The request appealed not only to Pasteur's scientific enquiring mind, but also to his intense patriotism. Even though Pasteur knew nothing of silkworms and had no idea that they suffered from disease, he accepted the challenge and visited Alés, the centre of the silk industry in the south of France, to begin his investigation. The outcome of his research was to forge another link in his 'inevitable' chain of discovery.

After nearly 18 months' study and a great deal of confusion, Pasteur discovered that, when examined under the microscope the silkmoth larvae were infected by bright oval bodies. He deduced that the silkmoth larvae were infested with parasites which they had picked up in the debris of the nurseries. After much deliberation, experimentation and confusion, Pasteur slowly realized, he was dealing with two distinct diseases, both of which depended upon the temperature, humidity and ventilation of the silkworm nurseries for their successful invasion.

The two different types of silkworm disease that Pasteur had identified were pebrine and flacherie. In pebrine black spots and corpuscles are generally, but not always, present on the worm. In such cases the worms often die within the cocoons. In the second type of disease, flacherie, the worms exhibit no corpuscles or spots but fail to spin cocoons. Pasteur suspected, but was not sure, that pebrine corpuscles were associated with the failure of the worms.

By examining the worms under the microscope, he was able to identify those that were free of pebrine and used only their eggs for breeding. Next he excluded from breeding eggs from worms with flacherie, which he identified by their sluggish behavior when climbing leaves to construct cocoons. Here lay the solution to the problem.

We now know that a member of the genus of parasitic spore-forming protozoans, called *Nosema*, of the order *Microsporida*, is responsible for the epidemic disease, pebrine. The other illness observed by Pasteur, flacherie, is of viral origin. The causative species of pebrine is called *Nosema bombycis*. It undergoes repeated divisions in host cells, followed by spore formation, and it attacks all tissues in all developmental stages, from embryo to adult.

Pasteur instructed the silkworm farmers on his methods of selection and on how to use a microscope to detect sickness in the worms. By adopting Pasteur's recommendation that the worms, moths and their eggs should be continuously sampled for signs of disease by microscopic examination, the French silkworm industry was slowly restored to profitability.

Pasteur's study of silkworm diseases came at a difficult period in his life as in 1868, he suffered a stroke which meant that he was paralyzed on his left side for the remainder of his life.

From then on, he had to rely upon assistants to perform many of his experiments for him, though his ability to use a microscope remained unimpaired.

Pasteur considered these studies to be important landmarks in his investigations on infection and infectious disease. As he expanded his research, he found that healthy worms became infected when allowed to nest on leaves used by infected worms. He also noted that the susceptibility of the worms varied widely: some worms died shortly after infection; some not until weeks later; some not at all. He determined that temperature, humidity, ventilation, quality of the food, sanitation and adequate separation of the broods of newly hatched worms each played a role in susceptibility to the disease. From this research, Pasteur began to formulate his concepts of the influence of environment on the spread of infectious disease.

INFECTIOUS DISEASE

Through Pasteur's ground-breaking discoveries, the Germ Theory which purported that certain diseases are caused by the invasion of the body by microorganisms – organisms too small to be seen except through a microscope – became more widely accepted. This discovery changed the whole face of pathology and effected a complete overhaul in the practice of surgery. Scientists and physicians throughout Europe began to build upon his successes.

The English surgeon Joseph Lister (1827–1912), in the 1860s for instance, believing from Pasteur's work that the reason why surgical wounds often turned septic was because of dangerous organisms in the hospital air, revolutionized surgical practice. By utilizing carbolic acid (phenol) as an antiseptic to exclude atmospheric germs, he was able to help prevent putrefaction in compound fractures of bones. In 1874, Pasteur himself declared that, if he had been a surgeon, he would 'never introduce an instrument in the human body without having passed it through boiling water'. Lister was so impressed by Pasteur's work that he began systematically to sterilize his instruments and bandages, and sprayed phenol solutions in his operating rooms, thus reducing infections following surgery to incredibly low numbers.

However, although the Germ Theory may have become largely accepted, its full implications for medical practice were not immediately apparent. Despite Lister's advocacy of the use of agents which prevent the growth of microorganisms, now known as antiseptics, Pasteur's idea of 'asepsis' – the removal of microorganisms altogether – took a surprisingly long time to be introduced into hospital practice. Bloodstained frock coats were considered suitable operating-room attire even in the late 1870s, and surgeons operated without masks or head coverings as late as the 1890s. As Pasteur wandered through hospital wards he became increasingly aware that infection was spread from sick to healthy patients by physicians and hospital attendants. Pasteur impressed on his physician colleagues that avoidance of microbes meant avoidance of infection. Slowly, but surely, through the preachings of Pasteur, Lister and other physicians, antiseptic medicine and surgery became the rule.

Within the next decade, identification and culture of the causative organisms of many serious infectious diseases, such as tuberculosis, cholera, diphtheria and tetanus, took place. Pasteur's rival, Robert Koch, was leading this field, largely as a result of his more reliable method of culturing bacteria on solid medium, while Pasteur turned his attention to the question of therapy.

✴ ✴ ✴ ✴SUMMARY ✴ ✴ ✴ ✴

- A better understanding of microorganisms and their role in disease and various 'chemical' processes brought forth an alternative theory; the Germ Theory.

- Pasteur spelt out his conviction that microorganisms were responsible for disease, putrefaction and fermentation.

- The development of the Germ Theory was an inspirational leap forward in our understanding of the role of microorganisms in infectious disease.

- Pasteur also contributed to the Germ Theory through an elegant study into the cause of disease in silkworms.

- The English surgeon Joseph Lister (1827–1912), believing from Pasteur's work that the reason why surgical wounds often turned septic was because of dangerous organisms in the hospital air, revolutionized surgical practice.

8 'Chance Favours the Prepared Mind'

PASTEUR'S CHANCE FIND

By the end of 1878 Pasteur's research had led him to experiment with the causative organism of fowl or chicken cholera, which was a serious problem for French farmers at that time. Chicken cholera, a disease in which the bird weakens, sleeps and dies, would spread through a farmyard rapidly and could decimate an entire flock in as little as three days. Pasteur's experiments were proceeding well; he had identified the cholera bacillus and was growing it in pure culture. He had additionally shown that the bacillus had little effect on guinea pigs, but was lethal to rabbits and when injected into chickens they invariably died in 48 hours.

During the heat of the summer of 1879 Pasteur and his staff took their well-earned annual holidays, leaving the cholera cultures used for infection stored on the shelves of the laboratory. Upon return, luck intervened. Pasteur found, not unexpectedly, that most of the cultures of chicken cholera had died off. Those that had not, he transferred to fresh culture infusions and tested them for potency by inoculation into healthy fowls. To Pasteur's surprise the chickens showed no disease symptoms. Presumably the old cultures no longer killed chickens, nor even made them sick. Nevertheless, to rule out other possibilities, the group set to work to make new cultures of the bacillus and tested these batches on the same fowls that had survived previously, as well as on a set of new birds. The results were astonishing. The previously injected birds were unaffected by the bacillus, while the new birds all died. To everyone's surprise and delight it appeared that whilst the old cultures of the chicken cholera had become weakened or attenuated, they actually provided protection to the birds from subsequent infection with fresh strong cultures of the disease.

Dismissing the idea that his attenuated cultures were an accidental discovery, Pasteur famously remarked, 'Chance favours the prepared mind.' He meant that he had been seeking an example of attenuation; his mind had, therefore, been prepared for any sign. Luck or not, these findings provided the breakthrough that led to development of specific vaccines to fight disease.

ATTENUATION OF GERMS

Pasteur grasped the opportunities that the fortuitous findings afforded, and proceeded to pinpoint the conditions required to produce successful attenuation of the chicken cholera cultures. Within only a few months, he had discovered that the secret of successful attenuation lay in exposing the cultures to air for a long period of time at an even temperature of about body temperature, 37°C. Pasteur realized that his findings, in a sense, echoed the studies of the English physician Edward Jenner (1749–1823) who, 80 years earlier, had conferred on humans immunity to smallpox by vaccinating individuals with a mild form of cowpox. Having revealed the conditions required to attain attenuation, Pasteur was able to reproducibly manufacture attenuated cultures of chicken cholera vaccines and could routinely prevent the disease in the vaccinated chickens.

ANTHRAX VACCINE

Flush with his success in producing a vaccine to protect chickens from fowl cholera, Pasteur turned his attention to anthrax, another disease which was reeking havoc in nineteenth-century French agriculture. Anthrax, a fatal disease of sheep and cattle, was decimating the sheep industry and the economy of France. Important strides in identifying the causative agent of anthrax, now classified as *Bacillus anthracis*, had been made by the time Pasteur entered the arena. The French physician Casimir-Joseph Davaine (1812–82) had, in 1863, previously observed bacteria in the blood of cattle which had died of anthrax and, by 1876, Koch had already isolated the anthrax bacillus from infected spleens. Koch had shown that, under resting conditions, the bacillus formed long-lived spores. Neither putrefaction nor heat killed them, and they could later develop into bacilli. These properties explained the persistence of the disease in fields. But there were still many aspects of the disease that needed explanation if outbreaks of the disease were to be controlled, infections cured and spread of the disease halted.

Anthrax is one of the oldest recorded diseases of animals, and in the eighteenth and nineteenth centuries it sometimes spread like a plague over the southern part of Europe, taking a heavy toll in human and animal lives. Practically all animals are susceptible to anthrax; cattle, sheep, goats, horses, and mules are the most commonly affected. But it was a puzzle in Pasteur's time as to how the disease spread.

Pasteur, during one of his excursions to a field where sheep were grazing, provided the answer. He noted that the ground in one part of the field was coloured differently compared to the rest. It was here that the farmer had buried some sheep which had died of anthrax, and the colour of the soil was due to earthworm casts. He realized that earthworms were feeding on the carcasses of the buried sheep and bringing the anthrax spores to the surface, where other animals could graze on the contaminated soil. It is now known that anthrax infection in non-grazing animals, such as swine, dogs, cats, and wild animals held in captivity, generally results from consumption of contaminated food.

The disease in animals is particularly unpleasant and may result in fever, spasms, respiratory or cardiac distress, trembling, staggering, convulsions, and death. Anthrax in humans occurs as a cutaneous, pulmonary, or intestinal infection:

* The most common presentation is as a primary localized infection of the skin in the form of a carbuncle. It usually results from handling infected material, lesions occurring mostly on the hands, arms, or neck. A small pimple develops rapidly into a large vesicle with a black necrotic centre, referred to as the 'malignant pustule'.

Should this condition become generalized, a fatal blood poisoning, or septicaemia, may ensue.

✳ The pulmonary form, called 'woolsorters' disease, affects principally the lungs and pleura, and results from inhaling anthrax spores in areas where hair and wool are processed. This form of the disease usually runs a rapid course and terminates fatally.

✳ The intestinal form of the disease, which sometimes follows the consumption of contaminated meat, is characterized by an acute inflammation of the intestinal tract, vomiting, and severe diarrhoea.

Anthrax is occasionally transmitted to humans by spore-contaminated animal-hair brushes or by wearing clothes such as furs and leather goods. Prompt diagnosis and early treatment are of great importance. Although containment of the animals on uncontaminated fields helped to control the spread of anthrax, Pasteur was well aware that more was needed.

He wondered whether, if attenuated cholera bacillus could render chickens resistant to that disease, an attenuated anthrax bacillus would render sheep immune to anthrax. The preparation of an anthrax vaccine was complicated by the need to prevent the bacteria turning into spores. This was achieved by attentuation at 42–44°C. By various techniques, involving oxidation and ageing, attenuated anthrax vaccines were made and they did indeed prevent anthrax in laboratory trials. Pasteur's reports on preventing sheep anthrax were so exciting to some and unbelievable to many, that he was challenged by a well-known veterinarian to conduct a carefully controlled public test of his anthrax vaccine.

This appealed greatly to Pasteur as he relished the publicity such a trial would bring. It was to take place before the Agricultural Society in 1881 at Pouilly le Fort, a farm in the town of Melun, south of Paris. Here, 25 sheep were to be controls, and another 25 were to be vaccinated by Pasteur. Then all the animals would receive a lethal dose of anthrax. For the trial to be called a success all of the control sheep must die and the vaccinated sheep must live.

When Pasteur's colleagues learned that he had agreed to the test they were concerned – a public success would be marvellous, but a public failure would be humiliating. The challenge was severe and there was no room for error. The vaccines were still in the developmental stage. The publicity was intense. A reporter from the *London Times* sent back daily dispatches. Newspapers in France followed the events with daily bulletins. There were crowds of onlookers, farmers, engineers, veterinarians, physicians, scientists and a carnival atmosphere. Would Pasteur's claims of vaccination hold up? Privately, even Pasteur was concerned that he had acted impetuously by accepting the challenge. Happily, the trial was a complete success. Two days after final inoculation, every one of the 25 control sheep was dead and every one of the 25 vaccinated sheep was alive and healthy.

These attenuated forms of disease-causing agents, following the smallpox vaccination model, became known as vaccines. Anthrax was the first infectious disease against which a *bacterial* vaccine was found to be effective. These discoveries laid the foundations for development of the modern science of immunology.

The fame of Pasteur and these experiments spread throughout France, Europe and beyond. From then on, Pasteur's laboratory began to 'manufacture' vaccines for sale to farmers and veterinary surgeons. Within ten years a total of 3.5 million sheep and 500,000 cattle had been vaccinated with a subsequent mortality of less than one per cent. The immediate savings to the French economy were enormous; estimated to be sufficient to cover the reparations that France was required to pay to Germany for the loss of the Franco–German War in 1880.

Today, anti-anthrax serum, together with arsenicals, and antibiotics, is used with excellent results. The hazard of infection to industrial workers can be reduced by sterilization of potentially contaminated material before handling, the wearing of protective clothing, use of respirators, and good sanitary facilities. Agricultural workers should avoid skinning or opening animals that have died of the disease.

RABIES VACCINE

The final and certainly most famous success resulting from Pasteur's research was the development of a vaccine against rabies, or hydrophobia as it is also known. Human rabies is the most severe of all communicable diseases. The infection usually results from the bite of a rabid dog or other mammal and, once infection is established is almost always fatal.

The association between animal bites and fatal human diseases has been recognized in many parts of the world for more than 2000 years. However, rabies has never been a major cause of human mortality and, unlike such epidemic diseases as smallpox, plague, and influenza, has had little impact on historical events. Some readers may wonder why such an eminent scientist as Pasteur chose a statistically unimportant human disease for investigation.

There are probably several reasons why Pasteur chose rabies as his next challenge, not least of which being that the disease has always had a hold on the public imagination and has been looked upon with horror. The possibility of finding a means of defeating such a disease may well have appealed to Pasteur's delight at 'playing to an audience'; he had a considerable flair for publicity.

The horror of rabies is understandable when the course of the disease is considered. Classically transmitted to its victim by the bite of a mad dog, a very long incubation period follows. This must be endured before the disease declares itself or the bitten person can be reasonably certain of having escaped infection. Finally, the disease itself is both terrifying and agonizing. Onset begins with a low fever, loss of appetite, headache, and often a recurrence of pain or tingling at the site of the bite. During the next few days there is growing anxiety, jumpiness, disorientation, neck stiffness, and sometimes epileptic seizures. Within a week the characteristic fear of swallowing starts. The patient may be consumed with thirst, but any attempt to drink induces violent spasms of the diaphragm, pharynx and larynx, with gagging, choking and a growing sense of panic; hence the term hydrophobia. As the condition worsens, even the sight or sound of water prompts these reactions and there are intervals of manical behaviour with thrashing, spitting, biting and raving. Delusions and hallucinations develop. These attacks alternate with periods of lucidity in which the patient suffers acute anxiety and mental distress. The nerves controlling eye movement and facial expression become paralyzed, and coma and death occur, usually within a week of the onset of the severe symptoms.

Attempts to treat and manage the symptoms of rabies, at a time when ignorance of the real nature of disease and how it spread incited fear and a sense of helplessness, were equally horrific. Bite wounds were often cauterized with a red-hot poker. Vinegar, salt and herb concoctions may have been used to wash the wounds, which would be deliberately kept open for as long as the patient lived. Patients were bled copiously every fortnight and purged every week with salts and purgative. Enemas of oil and induced vomiting were less common treatments.

Pasteur and his colleagues Charles Chamberland (1851–1908) and Pierre Emile Roux (1853–1933) realized that the conquest of rabies would be recognized as a great achievement in the world of science and by the public at large.

Pasteur would also, no doubt, have had childhood memories of rabies victims. The period of the Napoleonic wars, and the 20 or so years following them, coincided with a European outbreak of rabies. After December 1803, foxes were seriously involved in the progress of the epidemic amongst susceptible wild animals. Many foxes were infected in the Jura region of France, where Pasteur was born and spent his childhood. Foxes were found dead there and people, dogs, pigs and other animals were reportedly bitten. It was not until Pasteur was at least 13 years old, around 1835, that this particular European outbreak subsided.

Pasteur began his work on rabies at the beginning of the 1880s. He initially attempted to isolate the causative microorganism of rabies but, unsurprisingly his attempts were not successful. We know today that a virus causes rabies – viruses cannot be seen by light microscopy and cannot be grown on the culture media used for bacteria. Nevertheless, Pasteur tenaciously continued his research, attempting to transfer infection by injecting healthy dogs with saliva from rabid animals.

Pasteur was certainly courageous as he probably knew better than almost anyone how dangerous rabies was, how universally fatal and horrible the death. Yet he was willing literally to 'look into the jaws of death', as he and his colleagues faced the slavering jaws of a rabid dog held down by assistants so that samples of saliva could be obtained. Pasteur knew all the dangers but persisted.

It was not until some time later, that Pasteur had a new theory about rabies – that it must be an infection of the nervous system and that the disease had to get into a nerve. He acted on his hunch and found the active agent in the spinal cord and brain. By applying extracts of rabid spinal cord directly to the brain of dogs, he could reproducibly induce rabies in the test animals. Pasteur found that the brain tissues of dogs and rabbits provided an ideal culture medium for the disease.

Pasteur knew from his successes with chicken cholera and anthrax vaccines that, if it were possible, he had to find some way of producing an attenuation of the infectious agent of rabies. The first step was to get an animal to survive the injection. This seemed impossible. Pasteur knew of no person who had ever survived an attack of rabies. The germ was one of the most deadly known. This dangerous work went on for months as the team tried again and again. At last one dog developed the disease but, after a short illness, made a complete recovery. Excitingly Pasteur again injected the animal with the deadly rabies germs. The dog was immune. Pasteur now knew that vaccination against rabies was possible.

The next goal was to develop a vaccine that would provide protection to the subject before the rabies-causing agent moved from the bite site to the spinal cord and then to the brain.

By 1882, Pasteur had confirmed that rabies was an infection of the nervous system and that its virulence could be weakened by successive transfers through the spinal cords of some 25 rabbits.

This was achieved by injecting into test animals suspensions of spinal cord form rabid rabbits. These had been attenuated in strength by air drying over a 12-day period in the now-famous 'Roux' bottle. A strip of spinal cord was suspended from a hanger in the centre of the bottle. There was a hole at the top of the bottle and one on the lower side. Air entered from the bottom opening, passed over a drying agent and exited from the top. The longer the cord was dried, the less potent the tissue was in producing rabies. After successive transfers in the brains of rabbits, 'wild' virus from the saliva of rabid animals changed its character. The long incubation time of approximately a month in the natural infection became shortened to a fixed period of 5–8 days.

The treatment plan used to develop immunity to rabies was to inject under the skin of a dog the least potent preparation of minced spinal cord, followed every day for the next 12 days with a stronger and stronger extract. At the end of this time, the animal was completely resistant to bites of rabid dogs and failed to develop rabies if the most potent extracts were applied directly to the brain. This attenuated virus conferred immunity upon healthy dogs and offered a method of vaccination.

Following confirmation of his reports in 1885 that he had made dogs resistant to rabies by vaccination, Pasteur received wide acclaim and much favourable publicity. Not unexpectedly there were calls to test the vaccine on humans. But Pasteur was terribly afraid of things going wrong and he was particularly uneasy about being unable to isolate the causative organism. And so he continued to insist that many years of additional research were necessary before the treatment could be tried on humans. Pasteur had even considered trying the vaccine on himself before subjecting someone else to the treatment but was dissuaded by his colleagues.

In the event, a situation developed which made him act sooner. In July 1885, nine-year-old Joseph Meister and his mother appeared from Alsace at Pasteur's laboratory. Two days earlier a rabid dog had savagely mauled the young boy. His mother, having heard of Pasteur's work on animal rabies which by now had become quite well known, appealed to Pasteur to treat her son. All the indications were that the child had a month to live before the dreadful and fatal symptoms of hydrophobia would develop. After consultation with physician colleagues and with much trepidation Pasteur, having nothing to lose by the boy's death, began a course of vaccine therapy. Despite Pasteur's fears, it was soon clear that the symptoms of rabies were not going to appear, and Meister made a perfect recovery and remained in fine health for the remainder of his life.

A few months later, a second victim turned up. He was a young shepherd who had also been bitten by a mad dog. Following reports of these successful treatments, victims of dog and wolf bites from France, Russia, and the United States poured into his laboratory for treatment. The newspapers and public followed these treatments and cures with intense interest and the wild acclaim for Pasteur knew no bounds. Pasteur became a hero, and a legend in his own lifetime. To treat the increasing numbers of victims of rabies who were coming to Pasteur's laboratory an institute was built in Paris, funded by public and governmental subscriptions,. It was called the Pasteur Institute.

The year following the introduction of a rabies vaccination, some 2500 people were treated. World figures indicate that at the turn of the millennium approximately 1.5 million people were treated annually, most of them with vaccines which differ little from those prepared almost a century ago. Certainly, the Pasteur technique continued to be used in Paris until the 1950s, with all the associated problems of sourcing freshly infected rabbit cords, although these difficulties were partly resolved by storing cords in glycerine in which only minor losses of infectivity occurred. However, such difficulties, and possibly a reluctance to administer large quantities of even attenuated live rabies virus to humans, led workers in different parts of the world to modify Pasteur's vaccine.

Dilution of infective cords, rather than desiccation, was favoured by some; but most workers began to heat or treat chemically suspensions of infected nervous tissue – usually the whole brain rather than spinal cord – so that the virus was either partially or completely killed. The most notable contributor to this approach was Semple, an Englishman working in India. In 1911 he prepared a vaccine in which the virus was killed by incubating infected brain suspensions with carbolic acid. The bulk of the current world production of rabies vaccine is still prepared by Semple's method.

Thus, for almost a century, the infected neural tissues of adult animals have provided abundant sources of virus for the production of rabies vaccines and these notoriously crude preparations have presented the only possibility of preventing rabies infection from becoming fatal. The amount of inactivated virus that these preparations contain is several million times less than their content of brain tissue.

Pasteur's attenuated rabies virus is still used as starting material or 'seed virus' for most of the rabies vaccines prepared in the world today. It should also be noted that these classical researches by Pasteur and his co-workers established a model in practical immunology which has been exploited for preparing vaccines to control many other infectious diseases.

VACCINE THERAPY TODAY

Today vaccine therapy is one of the major strategies in the worldwide control of public health. Vaccines comprising of suspensions of either weakened or killed microorganisms that are capable of causing antibody production against an infectious microorganism when artificially introduced into the body, thereby conferring immunity from a subsequent infection of that microorganism, are now routinely administered. Vaccines consisting of the weakened toxins produced by microorganisms are also used. Once stimulated by a vaccine, the antibody-producing cells of the body remain sensitized to the infectious agent and respond to reinfection by producing more antibodies, thus reinstituting the immune response. Vaccines may be produced from either bacteria or viruses, although they have been most effective in preventing viral diseases. Vaccines of weakened, or attenuated, microorganisms generally produce a mild or subclinical form of the disease, while those comprising of deactivated or dead microorganisms produce no infection but are generally less potent. Attenuated vaccines used today include those for measles and hepatitis.

Smallpox has been eradicated through vaccination, and vaccines against polio, diphtheria, whooping cough, measles, and rubella have largely controlled these diseases in the developed world. Effective vaccines have also been developed for typhoid and paratyphoid fevers, cholera, plague, tuberculosis, undulant fever, tularemia, chronic staphylococcal and streptococcal infections, tetanus, influenza, yellow fever, some types of encephalitis, Rocky Mountain spotted fever, typhus, and hepatitis B, although some of these vaccines are used only in selected population groups at high risk of infection. Interest in bacterial vaccines slackened with the introduction of antibiotics in the mid-twentieth century, but vaccines remain important in the fight against many infectious diseases.

In the late twentieth century new types of vaccines were developed with the help of advanced laboratory techniques. Medical researchers became able to identify those biochemical components of a pathogen, or disease-causing microorganism, that stimulate the immune response to that organism in the body. Such a biochemical component can then be produced in the laboratory and subsequently administered to humans, upon whom it acts like any other vaccine but without risk of infection. An improvement on this approach, using recombinant DNA technology, is to splice the gene which codes for the production of that immunity-causing component into the DNA of an entirely different, and harmless, microorganism. The genetically altered microorganism is then injected into humans where it stimulates antibody production, both to itself and to the pathogen whose genes have been incorporated into it. This approach potentially enables the carrier organism to function as a live vaccine against several different diseases if it has received gene splices from the relevant disease-causing microorganisms.

✳✳✳✳SUMMARY ✳✳✳✳

- Pasteur pinpointed the conditions required to produce successful attenuation of the chicken cholera cultures.

- Pasteur was able to reproducibly manufacture attenuated cultures of chicken cholera vaccines and could routinely prevent the disease in the vaccinated chickens.

- Pasteur realized that earthworms were feeding on the carcasses of buried anthrax-infected sheep and bringing the anthrax spores to the surface.

- Anthrax was the first infectious disease against which a bacterial vaccine was found to be effective. These discoveries laid the foundations for development of the modern science of immunology.

- The final and certainly most famous success resulting from Pasteur's research was the development of a vaccine against rabies, or hydrophobia as it is also known.

- Pasteur found that the brain tissues of dogs and rabbits provided an ideal culture medium for the rabies.

- Pasteur confirmed that rabies was an infection of the nervous system and that its virulence could be weakened by successive transfers through the spinal cords of rabbits.

- To treat the increasing numbers of victims of rabies an institute was built in Paris, called the Pasteur Institute.

- Smallpox has been eradicated through vaccination, and vaccines against polio, diphtheria, whooping cough, measles, and rubella have largely controlled these diseases in the developed world.

The Pasteur Institute

9

PASTEUR'S OWN PRIVATE RESEARCH LABORATORY

On 1 March 1886, Pasteur presented the results of his rabies treatment trials to the Academy of Sciences and called for the creation of a rabies vaccine centre. Pasteur's laboratory was besieged by requests for rabies vaccine and hundreds of victims were brought to him from all over Europe. An extensive and international public drive for funds financed the construction of the Pasteur Institute, a private but State-approved institute recognized by the President of France, Jules Grévy, in 1887 and inaugurated by his successor Sadi Carnot in 1888. In accordance with Pasteur's wishes, the Institute was founded as a clinic for rabies treatment, a research establishment for infectious disease and a teaching centre. The 66-year-old scientist went on to dedicate the last seven years of his life to the Institute that still bears his name.

During this period, Pasteur also came to know the joys of fame and was honoured throughout the world with prestigious decorations. His work was continued and amplified throughout the world by his disciples, the Pasteurians.

The rabies treatment, the swansong of his career, was Pasteur's greatest triumph. Pasteur remained Director of the Institute until his death. He worked daily at his bench until September 1895, when he suffered a further stroke; he died on 22 September at Villeneuve-l'Étang, near Paris. He was given a State funeral and was buried in a magnificent tomb within the Pasteur Institute.

A statue outside the Pasteur Institute commemorates the Alsation shepherd boy, Joseph Meister, who was savagely mauled by a rabid dog but who was successfully treated by Pasteur in July 1885.

CENTRE OF EXCELLENCE

For over a century, the Pasteur Institute has been at the forefront of the battle against infectious diseases. This worldwide biomedical research organization, based in Paris, was first to isolate the AIDS virus in 1983. Over the years, it has been responsible for breakthrough discoveries that have enabled medical science to control such virulent diseases as diphtheria, tetanus, tuberculosis, poliomyelitis, influenza, yellow fever and plague. Since 1908, eight Pasteur Institute scientists have been awarded the Nobel Prize for Medicine or Physiology.

A private, non-profit organization, the Pasteur Institute has a unique history of accomplishment. Apart from the great achievements of Pasteur himself, during his lifetime his colleagues Émile Roux and Alexandre Yersin discovered how to treat diphtheria with antitoxins; Elie Metchnikoff received the Nobel Prize in 1908 for contributions to scientific understanding of the immune system and Jules Bordet received the prize in 1919 for his discoveries on immunity; Charles Nicolle received it in 1928 for unravelling the mystery of how typhus is transmitted.

A new age of preventative medicine was made possible by such developments from the Pasteur Institute as vaccines for tuberculosis, diphtheria, tetanus, yellow fever, poliomyelitis, and hepatitis B. Since World War II, Pasteurian researchers have focused on molecular biology. Their achievements were recognized in 1965, when the Nobel Prize was shared by François Jacob, Jacques Monod and André Lwoff for their work on the regulation of viruses.

Today, the Pasteur Institute is one of the world's leading research centres; it houses 100 research units and close to 2700 people, including 500 permanent scientists and another 600 scientists visiting from 70 countries annually. The Pasteur Institute is also a global network of 24 foreign institutes devoted to medical problems in developing nations; a graduate study centre and an epidemiological screening unit.

Research Centre

In addition to the isolation of HIV-I and HIV-2, in the recent past, researchers at the Pasteur Institute have developed a test for the early detection of colon cancer, produced a genetically engineered vaccine against hepatitis B and a rapid diagnostic test for the detection of the *helicobacter pylori* bacterium which is implicated in the formation of stomach ulcers. Other research in progress includes the study of cancer and, specifically, an investigation of the role of oncogenes, the identification of tumour markers for diagnostic tests and the development of new treatments. One area of particular interest is the study of human papilloma viruses and their role in cervical cancers. Researchers are currently focusing on the development of various vaccines against many diseases, including AIDS, malaria, dengue and the *Shigella* bacterium.

Teaching Centre

Since its founding, the Pasteur Institute has brought together scientists from many different disciplines for postgraduate study. Today, approximately 300 graduate students and 500 postdoctoral trainees from close to 40 different countries participate in postgraduate study programmes at the Institute. They include pharmacists and veterinarians, as well as doctors, chemists and other scientists.

Epidemiological Reference Centre

Strains of bacteria and viruses from many different countries are sent to the Institute's reference centre for identification. In addition to maintaining this vital epidemiological resource, the Institute serves as advisor to the French government and the World Health Organization (WHO) of the United Nations. Pasteur scientists also help to monitor epidemics and control outbreaks of infectious diseases throughout the world. These activities have created a close collaboration between the Institute and the US Centers for Disease Control and Prevention.

PASTEUR'S FINAL RESTING PLACE

Pasteur was buried, a national hero, by the French Government. His funeral was attended by thousands of people. His remains, initially interred in the Cathedral of Notre Dame, were transferred to a permanent elaborately-designed tomb within the Pasteur Institute. Among the mosaic tile decorations are two commemorating his research on rabies (a shepherd muzzling a dog) and anthrax (a contented flock of sheep).

In a tragic footnote to history, Joseph Meister – the first person publicly to receive the rabies vaccine – returned to the Pasteur Institute as an employee where he served for many years as gatekeeper. In 1940, 45 years after his treatment for rabies, which made medical history, he was ordered by the German occupiers of Paris to open Pasteur's crypt. Rather than comply, Joseph Meister committed suicide.

PASTEUR'S GENIUS

Pasteur's work is not simply the sum of his discoveries. It also represents the revolution of scientific methodology. Pasteur superimposed two indisputable rules of modern research: the freedom of creative imagination, necessarily subjected to rigorous experimentation. He would teach his disciples not to put forward anything that they could not prove by experimentation. Louis Pasteur was a humanist, always working towards the improvement of the human condition. He was a free-thinking man who never hesitated to take issue with the prevailing, yet false, ideas of his time.

He ascribed particular importance to the spread of knowledge and the applications of research. In the scientist's lifetime, Pasteurian theory and method were put into use well beyond the borders of France. Fully aware of the international importance of his work, Pasteur's disciples dispersed themselves to wherever their assistance was needed. In 1891, the first foreign Pasteur Institute was founded in Saigon, now Ho Chi Minh City, Vietnam, launching what was to become a vast international network of Pasteur Institutes. Because Pasteur changed the world forever, his homeland and the world have long considered him a benefactor of humanity.

Timeline

Events In World History		Events in the advancement in medical knowledge	
1789–94	French Revolution.	1796	Introduction by Edward Jenner of smallpox vaccination using material from cowpox.
1800–15	Napoleonic Wars.		
c. 1830	Industrialization of Europe begins.	1829	J.J. Lister invents the achromatic microscope.
1848	Riots in Paris, Vienna and parts of Germany and Italy.	1846	The anaesthetic ether is used for the first time in surgery. A year later chloroform anaesthesia is introduced.
1851–2	Louis Napoleon declares second French Republic.		
1854–6	Crimean War.	1854–6	Florence Nightingale becomes famous.
1861–5	American Civil War.	1859	Charles Darwin publishes *Origin of Species*.
1870	Franco-German War.	1865	Lister introduces antisepsis in surgery.
1871	Paris commune proclaimed. Third French Republic established.		
		1895	Discovery of X-rays.
1905	First Russian Revolution.	1897	Malaria is discovered to be transmitted by mosquitoes.
1914	World War I begins.	1909	Discovery of *Salvarsan* for the treatment of syphilis.

Significant outbreaks of disease		Events In The Life Of Pasteur	
		1822	Louis Pasteur born.
		1829–39	Attends school at Arbois.
1830–1	Influenza pandemic sweeps Europe.	1847	Awarded Doctor of Philosophy for his work on tartrate crystals.
1830–1	Cholera epidemic in Europe.		
1836–7	Influenza pandemic sweeps Europe.	1849	Marries Marie Laurent.
		1851	Son is born.
1840–60s	French Miliary Fever outbreaks.	1854	Dean of University of Lille.
1847–8	Influenza pandemic sweeps Europe.	1857–67	Director of Scientific Studies at the École Normale, Paris.
1848–9	Cholera epidemic in Europe.	1857	Announces Germ Theory of fermentation.
1853–4	Cholera epidemic in Europe.	1865	Starts studies on disease in silkworms.
1865–6	Cholera epidemic in Europe.	1868	Suffers first stroke.
1870s	European outbreak of smallpox.	1878	Starts studies on fowl cholera.
		1879	Finds and studies attenuation of germs.
		1881	Demonstrates an effective anthrax vaccine.
1887	French Miliary Fever outbreak.	1885	Demonstrates an effective rabies vaccine.
1889–90	Influenza pandemic sweeps Europe.	1895	Louis Pasteur dies.
1918–9	Pandemic of influenza sweeps Europe.		

Glossary

Abiogenesis The belief that living creatures arise from inorganic elements.

Aerobes Bacteria which are only active in air.

Alcoholic fermentation A process that converts glucose into alcohol and carbon dioxide.

Anaerobes Bacteria that are inactive in air.

Antibacterial Harmful to bacteria.

Attenuate Become weakened.

Bacilli Rod-shaped bacteria.

Bacteria (singular bacterium) A large and varied group of microorganisms, classified by their shape and staining ability. They live in many environments; only a few are pathogens.

Bacteriology The study of bacteria.

Biochemistry The study of the chemicals and their reactions in living organisms.

Contagious Infectious.

Crystallography The study of the structure and morphology of crystals.

Dextro-rotatory Describes a type of crystal or molecule that rotates to the right the direction of polarization of a beam of polarized light passing through it.

DNA (deoxyribonucleic acid) The main constituent of the chromosomes of all living organisms, except some viruses.

Enantiomer An optical isomer.

Enzyme A catalyst in a specific biochemical reaction.

Epidemic Large outbreak of a disease.

Fermentation The decomposition of carbohydrates by microorganisms, in the absence of oxygen.

Germs Microorganisms.

Germ Theory Theory of Pasteur that diseases are spread by living germs.

Goniometer A device for measuring angles and bearings.

Heterogenesis The appearance of living entities from a reconstruction of dead organic matter.

Hydrophobia Fear of water (condition present in rabies).

Immune Resistant to infection.

Immunology Study of the immune response.

Infectious disease Illness that can be passed on to others.

Isomers Chemical compounds of identical composition but differing in the arrangement of atoms within the molecule, and having different properties.

Laevo-rotatory Describes a type of crystal or molecule that rotates to the left the direction of polarization of a beam of polarized light passing through it.

Miasma A disease-causing vapour believed to be present in the air; the 'Miasmatic Theory' of disease was predominant up to the mid-nineteenth century.

Microbe Microorganism.

Microorgamism A living organism too small to be seen without a microscope.

Pandemic Epidemic covering a wide geographical area and affecting a large proportion of the population.

Pasteurization Technique of heating milk, beer, wine etc. to destroy bacteria it may contain; named after Pasteur.

Pathogen An organism that produces disease.

Polarized light Light in which the orientation of wave vibrations displays a definite pattern, for example being in a single plane.

Post mortem After death.

Protozoa Single-celled animals.

Putrefaction The process of rotting.

Racemic Describes a compound that contains equal amounts of dextro and laevo forms and is, therefore, optically inactive toward polarized light.

Recombinant DNA Any artificially created DNA that consists of a sequence from one genetic source spliced to a sequence from another.

Saprophytic organisms   Those microorganisms which neither cause harm nor good.

Splice   Technique used in gene cloning to add pieces of genetic material.

Spore   A dormant bacterium.

Spontaneous generation   A theory that living entities can arise either from inorganic elements (abiogenesis) or from a reconstitution of dead organic matter (heterogenesis) without a living 'parent'.

Staphylococci   Bacteria shaped like bunches of grapes.

Streptococci   Bacteria which aligns in chains.

Stereochemistry   The study of the spatial arrangement of atoms in molecules.

Symbiotic organisms   Those microorganisms which, whilst taking something from the body, give something in return.

Taxonomy   Classification into types or species.

Therapy   Course or type of treatment.

Toxin   A poisonous substance produced by a microorganism.

Vaccination, vaccine   Originally, the introduction of matter from cowpox pustules to lesson the danger of catching smallpox; by extension, vaccines are attenuated forms of disease used to confer immunity.

Virulent   Potent.

Virus   The smallest form of living animals, dependent on living cells for replication.

Further Reading

There are a very large number of books available which covers the work and life of Pasteur. The following brief list offers a few suggestions about where to begin further reading.

* *Dictionary of Scientists*, Oxford University Press, 1999.
  Entry on Louis Pasteur provides a concise biographical account of his life.

* *Super Scientists: the Silkworm Mystery – the Story of Louis Pasteur*, Pat Thomson, Hodder Wayland, 1998.
  The story tells how Louis Pasteur saved the French silkworm industry from doom.

* *Groundbreakers: Louis Pasteur*, Ann Fullick, Heinemann Library, 2000.
  A biography of Louis Pasteur. The book includes quotes and writings from newspapers and journals of the time.

* *Louis Pasteur*, Patrice Debre, The Johns Hopkins University Press, 2000.
  A detailed account of Louis Pasteur's life, struggles and contributions, written by a French immunologist and physician. The book draws heavily on Pasteur's own scientific notebooks and writings.

* *Louis Pasteur, Patrice Debre*, The Johns Hopkins University Press, 1998.
  Covers the bearing Pasteur's ideas have on the preoccupations of modern science, particularly with regards to his emphasis on the links that must exist between the patient's bed and the scientist's microscope. Pasteur's belief that basic research could not be totally separated from its practical applications was revolutionary in his age and continues to ignite debate today.

* *Louis Pasteur: Hunting Killer Germs*, EAM Jakab, McGraw-Hill Publishing Company, *Ideas on Trial*, 2000.
  The story of Pasteur from this pivotal moment in his career, through his amazing discovery of the fact that germs cause disease, to his development of Rabies and anthrax vaccines.

* *Louis Pasteur and Germs*, Steve Parker, Belitha Press, 1993.
  One of a series, which looks at the lives and discoveries of people in historical context

* *Pasteur and Modern Science*, Dubos, American Society of Microbiology (ASM), 1998.
  A biography of Louis Pasteur updated to the present day by microbiologist Thomas D Brock.

* *Pasteur's Fight Against Microbes*, Beverley Birch, Orion Children's Books *Science Stories*, 1995.
  Shows how Louis Pasteur's research led to the development of pasteurization and vaccination. This book has been written so that it can be used as support material for Key Stages 1 and 2 in National Curriculum Science.

* *The Private Science of Louis Pasteur*, Gerald L Geison, Princeton University Press, 1997.
  This text aims to penetrate the secrecy that has surrounded much of Pasteur's laboratory work. The author uses Pasteur's laboratory notebooks and published papers to present a detailed account of some of the most famous episodes in the history of science.

Websites

Here we have included some Internet web addresses for you to try out and surf amongst them at your leisure. We don't endorse or criticize any of them as they maybe just what you are looking for. We would suggest you try keying 'Louis Pasteur' into any search engine and seeing what results they turn up for you. Happy surfing.

http://www.pasteur.fr/pasteur/presentation/IP.html

http://www.louisville.edu/library/ekstrom/special/pasteur/cohn.html

http://ambafrance-ca.org/HYPERLAB/PEOPLE/_pasteur.html

http://encarta.msn.com/find/Concise.asp?ti=042F4000

http://home.inforamp.net/~schwartz/louis.htm

http://www.accessexcellence.org/AB/BC/Louis_Pasteur.html

http://www.invent.org/book/book-text/85.html

NEWTON –
A BEGINNER'S GUIDE

Jane Jakeman

Isaac Newton – A Beginner's Guide introduces you to the towering genius. Explore how his science revolutionized our world and his philosophy changed our thought. Find out more about Newton the man, and as scientist, philosopher, alchemist and respected public figure.

Jane Jakeman's lively text;

- describes Newton's background and the times he lived in
- explores his scientific ideas and their effect on our lives
- delves into the character of the man
- examines the influence of Newton on his own time and today.

The facts … the concepts … the ideas …

EINSTEIN – A BEGINNER'S GUIDE

Jim Breithaupt

Einstein – A Beginner's Guide introduces you to the great scientist and his work. No need to wrestle with difficult concepts as key ideas are presented in a clear and jargon-free way.

Jim Breithaupt's lively text:

- presents Einstein's work in historical context
- sets out the experimental evidence in support of Einstein's theories
- takes you through the theory of relativity, in simple terms
- describes the predictions from Einstein's theories on the future of the universe.

The facts ... the concepts ... the ideas ...

CHARLES DARWIN – A BEGINNER'S GUIDE

Gill Hands

Charles Darwin – A Beginner's Guide introduces you to the man whose scientific observations on evolution challenged the religious beliefs of Victorian society, but which are now generally accepted as being perfectly logical. Examine the historical perspective of evolution and the various philosophical questions that arise. No need to wrestle with difficult concepts as key ideas are presented in a clear jargon-free way.

Gill Hands' informative text explores:

- Darwin's background the times he lived in
- the development of the theory of natural selection
- the scientific basis for evolution
- the relevance of his ideas in today's world.

The facts … the concepts … the ideas …

FREUD –
A BEGINNER'S GUIDE

Ruth Berry

Freud – A Beginner's Guide introduces you to the 'father of psychoanalysis' and his work. No need to wrestle with difficult concepts as key ideas are presented in a clear and jargon-free way.

Ruth Berry's informative text explores:

- Freud's background and the times he lived in
- the development of psychoanalysis
- the ideas surrounding Freud's work on the unconscious.

The facts … the concepts … the ideas …

JUNG –
A BEGINNER'S GUIDE

Ruth Berry

Jung – A Beginner's Guide introduces you to the 'father of
analytical psychology' and his work. No need to wrestle
with difficult concepts as key ideas are presented in a clear
and jargon-free way.

Ruth Berry's lively text explores:

- Jung's background and the times he lived in
- the development of Jungian analysis in simple
  terms and key concepts and ideas surrounding
  his work
- the study of dreams and their interpretation
- the archetypal interpretations of popular myths
  and legends.
- the concept of the symbol

The facts … the concepts … the ideas …

SARTRE –
A BEGINNER'S GUIDE

George Myerson

Sartre – A Beginner's Guide introduces you to the life and work of this leading novelist, central philosopher and major dramatist. In the new millennium, Sartre remains a symbol of the committed writer and thinker, and his existentialism continues to challenge us. No need to wrestle with difficult concepts as key themes and ideas are presented in a clear jargon-free way.

George Myerson's fascinating introduction:

- brings the different phases of Sartre's thought and art to life
- explains the key ideas of Sartre's exiestentialism using examples from his work
- summarises essential information about characters, plots and arguments in the major works
- puts Sartre in philosophical and historical context.

The facts … the concepts … the ideas …

MARX –
A BEGINNER'S GUIDE

Gill Hands

Marx – a beginner's guide gives you the essential facts
surrounding the 'father of communism'. No need to wrestle
with difficult concepts as key themes and ideas are presented
in a clear and jargon-free way.

Gill Hand's no-nonsense text takes you step-by-step through:

- Marx's background and the times he lived in
- The ideas that led to revolutions throughout the
  world
- The place of Marxism after Marx

The facts … the concepts … the ideas …

DA VINCI –
A BEGINNER'S GUIDE

Ruth Berry

Da Vinci – A Beginner's Guide introduces you to the life and work of a great genius. Leonardo is usually thought of as an artist, but he was also an intellectual giant in the developing field of science and an accomplished musician, architect and engineer. Follow the story of a true genius rich in ideas.

Ruth Berry's lively text investigates:

- Leonardo's background and the times he lived in
- The importance of the Renaissance
- Leonardo's influence on the workd of art
- His astounding exporations in science and technology.

The facts … the concepts … the ideas …